CONNECTED INTELLIGENCE

AI and the Internet of Things

by
Charlie Morgan

CONNECTED INTELLIGENCE

AI and the Internet of Things

CONTENTS

INTRODUCTION

Imagine a world where everyday objects are not only connected to the internet but also imbued with the intelligence to make decisions, learn from experience, and adapt to our ever-evolving needs. This is not a vision of some distant future but a reality unfolding around us every single day. The integration of Artificial Intelligence (AI) and the Internet of Things (IoT) is transforming the landscape of technology, revolutionizing industries, and reshaping our daily lives in ways we could have never anticipated.

Artificial Intelligence, with its capacity to mimic human thinking and decision-making processes, is no longer confined to the realm of science fiction. It's here, permeating various facets of our existence, from virtual assistants guiding us through our schedules to complex algorithms diagnosing diseases with unprecedented accuracy. On the other hand, IoT connects ordinary devices to the internet, enabling them to send and receive data, which broadens their utility exponentially.

Now, envision an extraordinary synergy where AI and IoT come together. This convergence, often referred to as AIoT, is more than the sum of its parts. AI supercharges IoT devices, making them not only interconnected but also intelligent. These smart devices can predict, recommend, and even take preemptive actions, making our lives more convenient, efficient, and fascinating.

The aim of this book is to delve deep into this integration, to unravel the complexities, and to illuminate the vast potential of AIoT. As

1

we embark on this journey, we'll explore how AIoT is already making a tangible impact across various sectors—be it in smart homes, where AI-powered systems bring unprecedented convenience, or in healthcare, where intelligent monitoring can save lives. We'll look at industrial applications where predictive maintenance is not just saving money but also preventing disasters.

We're standing on the brink of a new era, where AIoT is redefining what we know about our surroundings and how we interact with them. In urban planning, for instance, AIoT enables more efficient traffic management systems, enhancing public safety and improving the quality of life in cities. In agriculture, precision farming powered by AI and IoT is maximizing yields while minimizing waste. These advancements are not siloed but interconnected, creating a fabric of intelligent solutions that redefine modern life.

It's essential to grasp the fundamentals before diving headfirst into the myriad applications and possibilities of AIoT. Understanding the core concepts of AI and IoT will provide a solid foundation upon which to build our exploration. We'll take you through the fundamental principles, the technological advancements that have paved the way, and the transformative power unleashed by the combination of these two fields.

Moreover, this book will illuminate the ethical, privacy, and security concerns that come with the territory. Are we ready to hand over such crucial aspects of our lives to intelligent systems? How do we ensure that our data is secure and our privacy respected? What regulatory frameworks are evolving to keep pace with these rapid advancements? We'll address these pressing questions, helping you navigate the landscape of AIoT with a balanced perspective.

Businesses are also at a crossroads. The implementation of AIoT solutions requires strategic planning, investments, and an understanding of the market dynamics. From enhancing customer experiences in retail to optimizing supply chains, AIoT offers a competitive edge that is hard to ignore. This book will provide insights into business strategies, helping companies leverage AIoT for sustainable growth and profitability.

Looking towards the future, the possibilities seem boundless. Emerging trends like quantum computing and edge AI promise to propel AIoT to even greater heights. Imagine autonomous vehicles that not only drive themselves but learn from every trip, optimizing routes in real-time and communicating with other smart vehicles to prevent accidents. Picture smart cities where resources are managed so efficiently that energy waste is a thing of the past.

Education, entertainment, energy management, finance—no sector remains untouched by the AIoT revolution. We'll explore how smart classrooms facilitate personalized learning, how AI-driven entertainment systems curate content tailored to our preferences, and how intelligent grids are making energy distribution more efficient.

The environmental impact of AIoT cannot be overstated. As we face pressing challenges like climate change and resource depletion, AIoT offers innovative solutions for monitoring and managing our environmental footprint. From intelligent waste management systems to AI-driven climate models, the potential for positive impact is immense.

This book is not just a technical guide but also a source of inspiration and motivation. We'll highlight real-world applications, case studies, and success stories that demonstrate the transformative power of AIoT. These narratives will not only inform but also inspire, showcasing how visionaries and innovators are harnessing AIoT to solve complex problems and drive societal progress.

In conclusion, the integration of AI and IoT is more than a technological trend; it's a paradigm shift that is redefining our world. This book is your gateway to understanding this transformation, its implications, and the vast potential it holds for the future. Whether you're a tech enthusiast, a business leader, or simply curious about the world of AIoT, this journey promises to offer profound insights, practical knowledge, and a glimpse into the exciting future that's unfolding right before our eyes.

CHAPTER 1:
THE EVOLUTION OF AI AND IOT

The intersection of Artificial Intelligence (AI) and the Internet of Things (IoT) marks a significant shift in technological advancement, a fusion that continues to transform industries and reshape our daily lives. AI, with its capabilities of simulating human intelligence and learning, combined with IoT, which connects everyday objects to the internet, creates a powerful ecosystem known as AIoT. This convergence has unleashed a wave of innovation, allowing devices to not only gather data but also to analyze and respond to it in real-time, driving efficiency and opening doors to new possibilities. Over the past decades, we've witnessed how these two technologies have evolved individually—from the early days of machine learning algorithms and networked devices to today's sophisticated AI models and interconnected smart environments—culminating in a synergy that is more impactful than the sum of its parts. As industries adopt AIoT solutions, we're seeing smarter cities, advanced healthcare systems, and streamlined industrial operations, all paving the way for a future where intelligent systems intuitively enhance our lives and work environments.

Understanding Artificial Intelligence

Artificial Intelligence (AI) has rapidly evolved from the realms of science fiction to an integral part of modern technology and daily life. At its core, AI refers to the simulation of human intelligence in machines programmed to think, learn, and make decisions much like the human brain. This involves various subfields such as machine learning, natural language processing, and robotics.

The roots of AI can be traced back to early attempts at creating machines that could perform tasks requiring human intelligence, such as calculation and data processing. However, the modern era of AI began in the mid-20th century with pioneers like Alan Turing and John McCarthy, who laid the theoretical foundations for the development of AI algorithms and systems. The advent of digital computing provided the necessary hardware to bring these theories to life, resulting in the first AI programs that could solve mathematical problems and play simple games.

Machine learning, a subset of AI, has been particularly transformative. It enables computers to learn from data and improve their performance over time without being explicitly programmed. This process resembles the way humans learn from experience. For instance, through supervised learning, algorithms are trained on labeled datasets and can subsequently make predictions or decisions based on new, unseen data. Techniques like neural networks and deep learning have further enhanced the capabilities of AI, allowing it to process vast amounts of data and recognize patterns with astounding accuracy.

A significant milestone in AI development was the creation of neural networks modeled after the human brain's structure. While early neural networks were relatively simple, advancements in processing power and data collection have led to the development of deep learning models with many layers of interconnected neurons. These models excel in complex tasks, such as image and voice recognition, and have been instrumental in enabling applications like virtual assistants and autonomous vehicles.

Natural language processing (NLP) is another compelling application of AI, focusing on the interaction between computers and human language. NLP facilitates applications ranging from voice-activated assistants like Siri and Alexa to sophisticated chatbots that can understand and respond to human queries. This has revolutionized how we interact with technology, making interfaces more intuitive and accessible.

In recent years, AI has moved from theoretical research and academic labs to practical, real-world applications across various industries. In healthcare, AI algorithms analyze medical data to aid in diagnostics and treatment planning. In finance, AI models are employed for algorithmic trading and fraud detection. The retail sector uses AI for personalized recommendations and inventory management, ensuring a tailored shopping experience that meets consumers' unique preferences.

AI's potential extends even further when combined with the Internet of Things (IoT). IoT refers to the network of interconnected devices that collect and exchange data in real time. By embedding AI into IoT devices, we can create smart systems that not only collect data but also analyze and act upon it autonomously. This convergence, often termed AIoT, amplifies the capabilities of both AI and IoT, leading to innovations like smart homes, intelligent transportation systems, and advanced manufacturing processes.

The synergy between AI and IoT can be witnessed in smart homes, where devices like thermostats, lighting systems, and security cameras communicate with each other and make intelligent decisions based on user behavior and preferences. This not only enhances convenience but also promotes energy efficiency and security. In industrial settings, AI-powered IoT systems monitor machinery and optimize operations, reducing downtime and maintenance costs. Predictive analytics enabled by AI helps in foreseeing potential machine failures and addressing them before they occur.

Educational institutions are also harnessing the power of AI to create personalized learning experiences. AI algorithms analyze student performance data to identify learning gaps and suggest tailored resources. This individualized approach fosters a more effective learning environment, catering to diverse student needs. Furthermore, AI-driven administrative systems streamline operations, allowing educators to focus more on teaching and less on bureaucratic tasks.

Despite the immense potential and benefits of AI, it's essential to recognize and address the ethical considerations and challenges associated with its deployment. Issues such as data privacy, algorithmic bias, and transparency are critical to ensuring that AI systems are fair and trustworthy. Regulatory frameworks and ethical guidelines must evolve in tandem with technological advancements to safeguard users' rights and promote responsible AI usage.

In summary, understanding artificial intelligence involves comprehending its foundational principles, diverse applications, and the profound impact it has on various aspects of our lives. As AI continues to evolve and integrate with IoT, we stand at the cusp of a technological revolution that promises to transform industries, enhance daily life, and shape the future in ways we are only beginning to imagine.

The Development of the Internet of Things

The Internet of Things (IoT) has fundamentally reshaped the technological landscape, creating a world where everyday objects are interconnected and communicate seamlessly. While the term "Internet of Things" itself was coined in the late 1990s, the roots of this concept can be traced back to early networked systems and sensor technologies developed in the 20th century. In essence, IoT refers to the network of physical devices that collect and exchange data, leveraging the power of the internet. It's the smart home systems we use, the wearable devices that monitor our health, and the industrial machines that optimize manufacturing processes.

IoT's evolution has been driven by several key technological advancements. Early steps were taken with the development of RFID (Radio Frequency Identification) technology, initially used for tracking goods in supply chains. As wireless networks and general internet accessibility expanded, so did the concept of interconnected devices. Progress in miniaturization, wireless communication, and sensor technology paved the way for the integration of these systems into everyday items, from household appliances to complex industrial equipment.

Parallel to these advancements, software development and data analytics grew more sophisticated. The enormous volumes of data generated by IoT devices required new approaches to storage, processing, and analysis. Cloud computing emerged as a critical component, providing on-demand computational resources and storage capacity crucial for handling IoT data streams. Edge computing, which processes data closer to where it is generated rather than relying solely on centralized cloud servers, further minimized latency and improved the efficiency of IoT operations.

In essence, IoT interconnects myriad devices, making them "smart" by embedding sensors and actuators. These smart devices range from simple household items like smart thermostats and light bulbs to advanced industrial machinery. One key characteristic of IoT devices is their ability to collect real-time data, which can then be analyzed to derive actionable insights. This real-time data collection allows for predictive maintenance in industrial settings, personalized health monitoring in healthcare, and more.

One striking aspect of IoT's development is its impact on various sectors. In healthcare, for example, IoT has enabled the creation of advanced patient monitoring systems and personalized care solutions. Wearable devices track vital signs and transmit data to healthcare providers, facilitating continuous monitoring and timely interventions.

The energy sector has benefited from smart grids, which optimize electricity distribution and reduce waste. Meanwhile, in agriculture, IoT sensors monitor soil conditions, crop health, and weather patterns to enhance yield and sustainability.

Despite these advancements, the development of IoT has not been without challenges. Security and privacy concerns are paramount, given the vast amount of data IoT devices generate and transmit. Each connected device represents a potential vulnerability, and the proliferation of these devices has heightened the risk of cyberattacks. Strong security protocols and encryption are essential, but so too is a robust framework for regulating and managing these technologies to protect user data and privacy.

Moreover, IoT has catalyzed significant shifts in business models and consumer behavior. Companies now leverage IoT data to offer personalized services and products, creating value in new and innovative ways. The proliferation of smart devices has also led to an increased demand for seamless user experiences and intuitive interfaces. As IoT continues to evolve, the focus on user-centric design and accessibility will remain crucial to its widespread adoption.

The convergence of IoT with artificial intelligence (AI) has further propelled its growth. AI's ability to analyze complex data sets and derive meaningful insights complements IoT's data generation capabilities. Together, they create systems that are not only interconnected but also intelligent. For instance, AI-powered algorithms can process IoT data to predict equipment failures in manufacturing, enhancing operational efficiency and reducing downtime. Similarly, AI-driven analytics can optimize energy consumption in smart homes, leading to more sustainable living practices.

The impact of IoT extends beyond individual sectors to encompass broader societal implications. Smart cities, for example, leverage IoT to improve urban infrastructure and services. Sensors embedded throughout the city monitor traffic patterns, air quality, and energy usage, providing city planners with real-time data to enhance public safety, reduce congestion, and promote sustainability. These smart city initiatives exemplify the transformative potential of IoT on a societal level.

Looking ahead, the continued development of IoT will undoubtedly bring about further technological advancements and societal changes. The integration of 5G technology promises to enhance the speed and reliability of IoT networks, enabling even more devices to communicate seamlessly. Additionally, advances in artificial intelligence, machine learning, and data analytics will further enhance the capabilities of IoT systems, driving innovation across various domains.

In conclusion, the development of the Internet of Things has been a multifaceted journey, marked by significant technological advancements and widespread societal impact. From its early roots in RFID technology to its current status as a cornerstone of modern technology, IoT has transformed how we interact with the world around us. As we move forward, the synergy between IoT and AI will continue to drive innovation, creating smarter, more efficient systems and unlocking new possibilities for technological and societal progress.

The Convergence of AI and IoT

In examining the convergence of Artificial Intelligence (AI) and the Internet of Things (IoT), one must first foundationally understand their intrinsic qualities. Individually, AI and IoT have profoundly transformed how we interact with technology and our environments. Together, they are driving a technological revolution that touches every industry and aspect of daily life.

The fusion of AI and IoT, often termed "AIoT," epitomizes the technological symbiosis where AI's computational intelligence leverages the vast data flow from IoT devices. This amalgamation empowers systems to not just collect and analyze data, but to autonomously make decisions and derive actionable insights. Consider it as a confluence where real-time data gathering meets advanced analytics and decision-making.

IoT devices by themselves are marvels of modern engineering, creating an intricate web of connectivity across various touchpoints. Smart sensors, embedded systems, and communication protocols weave a digital fabric that extends to homes, vehicles, cities, and industries. AI enhances this fabric by infusing it with learning algorithms, predictive analytics, and natural language processing capabilities. This intelligent weaving results in systems that anticipate actions, respond to inquiries, and learn from every interaction.

Imagine a smart city where AI algorithms manage traffic flows, emergency response, waste collection, and even air quality. IoT sensors collect data from nearly every corner, allowing AI systems to analyze and optimize urban functionalities in real-time. This transformation is not confined to urban settings. In the healthcare sector, AIoT is accelerating the shift from reactive to preventive care. Wearable devices and remote monitoring systems powered by AI can alert healthcare providers to anomalies before they escalate into critical health issues.

Beyond individual use cases, the convergence of AI and IoT is driving efficiency and innovation in industries such as manufacturing. Here, predictive maintenance powered by AI analytics on IoT sensor data preempts equipment failures, reducing downtime and saving costs. AI also optimizes manufacturing processes by analyzing workflow data collected by IoT devices to recommend improvements. This robust synergy enables a dynamic, continuously improving operational landscape.

Retail, an industry characterized by rapid evolution and fierce competition, leverages AIoT to enhance customer experiences and streamline operations. Real-time inventory tracking through IoT combined with AI-driven demand forecasting ensures products are always available without overstocking. AI algorithms personalize marketing efforts based on data collected from in-store sensors, loyalty programs, and online behavior—creating a seamless, individualized shopping experience.

Transportation and logistics are similarly revolutionized. Autonomous vehicles collect vast amounts of data via IoT devices, processed by AI to navigate, respond to traffic conditions, and maintain optimal performance. In logistics, AI-driven analytics on data from IoT-enabled fleets enhance route optimization and fuel efficiency, reducing costs and environmental impact.

Educational technologies, or EdTech, also benefit from this convergence. Smart classrooms equipped with IoT devices can monitor student engagement and environmental conditions. AI analyzes this data to tailor educational content to individual learning styles and improve administrative efficiencies. This adaptive learning environment aims to foster more personalized and effective education.

The collaborative potential between AI and IoT extends to environmental initiatives as well. IoT sensors track environmental parameters across vast geographical areas, while AI models analyze this data to identify patterns and predict climate changes. Smart grids and renewable energy management systems use AIoT to optimize resource allocation, enhancing sustainability efforts without sacrificing efficiency.

While the prospects are promising, the hurdles remain significant. Technical challenges such as data security, privacy concerns, and interoperability between different IoT devices and AI platforms need addressing. However, the amalgamation of AI and IoT continues to push these boundaries, driving innovation that promises more connected, intelligent, and efficient systems globally.

This convergence isn't just about technology but also about creating a transformative impact on society. The way we live, work, and interact with our environments is undergoing a significant shift. The potential of AIoT is boundless, limited only by the creativity and ingenuity applied to harnessing its power.

Embracing the convergence of AI and IoT necessitates a mindset willing to partner with technology to solve complex problems. It calls for a readiness to innovate and adapt continually. As we move further into the 21st century, this fusion will undoubtedly be a cornerstone of technological advancement, reshaping industries and redefining possibilities.

Ultimately, AIoT is not just a technological innovation but a testament to human ambition and ingenuity. It is a step toward a future where intelligent systems work harmoniously with us, improving livelihoods, industries, and the overall quality of life. As we harness AIoT's combined potential, we're not merely creating smarter technologies but fostering an interconnected, intelligent world where the potential for progress and innovation is limitless.

CHAPTER 2:
FUNDAMENTALS OF AI AND IOT

The intersection of Artificial Intelligence (AI) and the Internet of Things (IoT) establishes a transformative framework that is reshaping industries and redefining our everyday experiences. AI, with its data-driven algorithms and predictive capabilities, enhances the functionality of IoT devices by enabling them to learn from data, adapt to new inputs, and autonomously make informed decisions. Similarly, IoT provides the essential data infrastructure and connectivity that AI systems depend on for real-time insights and actions. This synergy drives innovation across various sectors, from smart homes improving daily conveniences to industrial IoT systems optimizing manufacturing processes. Together, AI and IoT create an ecosystem where devices are not only interconnected but also intelligent, leading to unprecedented efficiency, personalization, and automation.

Core Concepts of AI

Artificial Intelligence (AI) is a transformative pillar within the larger framework of modern technology, fundamentally altering how we interact with digital systems, data, and even each other. In its essence, AI is the simulation of human intelligence processes by machines, especially computer systems. These processes include learning (the acquisition of information and rules for using that information), reasoning (using rules to reach approximate or definite conclusions), and self-correction. Broadly speaking, AI aims to create systems that can function autonomously, solving problems, making decisions, and performing tasks in a way that can rival or exceed human capabilities.

Machine Learning (ML) is a core component of AI. It's a method of data analysis that automates the building of analytical models. Using algorithms that iteratively learn from data, ML allows computers to find hidden insights without being explicitly programmed where to look. Deep learning, a subset of ML, leverages neural networks with many layers (hence "deep"), which is incredibly effective in handling large volumes of unstructured data like images, videos, and texts.

Another essential concept within AI is natural language processing (NLP). NLP is about the interaction between computers and humans using natural language. It encompasses everything from speech recognition and translation to sentiment analysis and content generation. The capacity of NLP to understand and generate human language is transforming industries like customer service, healthcare, and media.

Computer vision is another key area. It involves training machines to interpret and make decisions based on visual data from the world. It's utilized in various applications, such as facial recognition, medical imaging, and even autonomous vehicles. The advancements in this field have provided new ways to interpret and respond to the environment, which can be accurately measured and analyzed in real-time.

Expert systems, one of the earlier successful forms of AI, are designed to solve complex problems by reasoning through bodies of knowledge, represented mainly as if-then rules rather than through conventional procedural code. They are used in applications like medical diagnosis, equipment repair, and financial services.

Robotics, often seen as the physical manifestation of AI, integrates AI to perform tasks ranging from simple, repetitive actions to complex, human-like interactions and maneuvers. Advances in robotics, combined with AI, have led to the development of intelligent agents capable of decision-making and learning from their environments.

Reinforcement learning, another fascinating aspect, involves an agent continually improving its performance by receiving rewards or penalties from its actions and learning from its mistakes. This type of learning is instrumental in developing systems that optimize for long-term goals, such as autonomous driving or game-playing algorithms.

AI ethics plays a pivotal role in steering the development and deployment of AI technologies. It encompasses concerns about fairness, accountability, and transparency. The field is dedicated to ensuring that AI systems are used responsibly and that their decisions can be understood and trusted by humans. Ethical AI also emphasizes the mitigation of biases that could lead to discriminatory outcomes.

In terms of practical application, AI has found its way into various sectors like finance, healthcare, education, and even entertainment. Financial institutions use AI for credit scoring, fraud detection, and wealth management. In healthcare, it's revolutionizing diagnostics, personalized medicine, and patient care management. Educational platforms utilize AI for adaptive learning, offering personalized content based on students' learning paces and styles.

The integration of AI with the Internet of Things (IoT) is referred to as Artificial Intelligence of Things (AIoT), and it represents a powerful synergy that enhances the capabilities of interconnected devices. AI can enhance IoT systems by providing intelligent insights drawn from data collected from the network of connected devices, thereby optimizing performance, predicting maintenance needs, and enhancing user experiences.

The scope of AI is immense, and as we continue to explore its depths, the potential applications seem endless. From enhancing daily conveniences to delivering life-saving capabilities, AI's core concepts form the backbone of modern technological advancements. As AI continues to evolve, it holds the promise of not only solving today's problems but also transforming our future in ways we are only beginning to imagine.

Key Components of IoT

In the ever-evolving symphony of technology, the Internet of Things (IoT) stands as a crucial movement in the composition. At its core, IoT refers to a sprawling network of interconnected devices, all communicating seamlessly to perform various functions and enhance human life in multiple facets. But this marvel doesn't work by magic alone; it relies heavily on several key components that ensure its smooth operation and utility.

The first fundamental component of IoT is the *devices* or *sensors* themselves. These are the physical entities embedded with sensors, software, and other technologies to collect and exchange data with other devices and systems over the Internet. From simple temperature sensors to complex smartphones, the array of devices is vast and varied. Each one is designed to capture specific data about its environment or the user, sending this information to a central system for processing and decision-making.

Connectivity is the lifeline of any IoT infrastructure. Without a robust means to transfer data, IoT devices would be no more than isolated gadgets. This is where network protocols and communication technologies come into play. These include Wi-Fi, Bluetooth, Zigbee, LoRaWAN, and cellular networks. Each type of connectivity has its unique strengths and weaknesses, making it suitable for different applications. For instance, Wi-Fi offers high data rates but limited range, while LoRaWAN provides extensive coverage at the expense of lower data throughput.

Another critical component is the *gateway*. Acting as the bridge between IoT devices and the cloud or central processing units, gateways are responsible for transforming the data gathered by sensors into formats that can be understood and processed by the cloud. They often include security features to encrypt data and protect it from unauthorized access, making them vital for maintaining the integrity and confidentiality of the information being transmitted.

Given the massive amounts of data generated by IoT devices, *data processing* becomes another pivotal aspect. This processing can occur either at the edge—closer to the IoT devices—or in the cloud. Edge computing reduces latency and minimizes bandwidth usage by processing data locally and sending only essential information to the cloud. On the other hand, cloud computing offers virtually limitless processing power and storage, allowing for more complex analytics and long-term data retention. Balancing the two to achieve optimal performance and efficiency is often a key design consideration in IoT systems.

Security cannot be overlooked when discussing IoT. The extensive network of connected devices presents multiple entry points for cyber threats. Thus, *IoT security* measures must be implemented across various stages, from device manufacturing to data transmission and storage. Encryption, multi-factor authentication, intrusion detection systems, and regular software updates are some of the common security practices to safeguard the IoT ecosystem from malicious activities.

IoT platforms serve as the backbone for managing IoT networks. These platforms enable device management, data collection, integration with other systems, and analytics. They often come with dashboards and reporting tools to help users monitor performance, track anomalies, and make informed decisions. Popular IoT platforms include AWS IoT, Google Cloud IoT, and Microsoft Azure IoT, each offering a suite of tools tailored to different needs and applications.

Another essential component is *analytics*. Harnessing the power of data collected by IoT devices, analytics involve a range of techniques from simple statistical methods to advanced machine learning algorithms. These analytical tools allow for extracting meaningful insights, predicting future trends, and facilitating proactive decision-making. In industries like healthcare, predictive analytics can save lives by alerting medical practitioners to potential health risks before they become critical.

Application and service interfaces are the final piece of the puzzle. These interfaces enable end-users to interact with the IoT system easily. Whether it's a mobile app that controls smart home devices or a web dashboard that monitors industrial equipment, a well-designed interface makes the system accessible and user-friendly. APIs (Application Programming Interfaces) often play a crucial role in this, allowing different software applications to communicate with each other seamlessly.

The true power of IoT lies in the seamless blend of these components, creating an ecosystem where devices, connectivity, gateways, data processing, security, platforms, analytics, and interfaces work in unison. By understanding and integrating these key components, we unlock new possibilities and drive innovation, making our world smarter, more efficient, and interconnected.

Now that we've laid out the foundational elements of IoT, it's time to pivot and explore how Artificial Intelligence (AI) significantly enhances these systems. The integration of AI into IoT, often termed AIoT, extends the capabilities of IoT devices beyond mere data collection and sharing, ushering in an era of unprecedented automation, intelligence, and adaptability.

How AI Enhances IoT

In the rapidly evolving landscape of technology, the blend of Artificial Intelligence (AI) and the Internet of Things (IoT) is leading to groundbreaking developments. These two pillars of modern tech are forging a powerful symbiosis that's transforming industries and everyday life. At its core, AI adds cognitive capabilities to IoT devices, enabling them to not just collect data, but learn from it and make autonomous decisions. This transformation from passive to active interaction is the essence of how AI enhances IoT.

The fusion of AI and IoT, often referred to as AIoT, leverages AI's prowess in data analytics to handle the colossal amounts of data generated by IoT devices. Think of millions of sensors embedded in smart cities, homes, and industries, all producing data every second. Managing and extracting meaningful insights from this data is challenging, but AI's advanced algorithms, such as machine learning and neural networks, are designed precisely for such tasks. These algorithms don't just streamline the processing of data; they identify patterns and anomalies that humans might overlook, leading to smarter and faster decision-making processes.

Take smart homes as an example. Integrating AI into IoT-enabled devices enhances their functionality exponentially. AI algorithms analyze user behaviors, preferences, and routines to optimize the functioning of devices such as thermostats, lights, and security systems. This results not only in heightened user convenience but also in significant energy savings. AI can predict when you're likely to be home and adjust the home's environment accordingly, ensuring comfort while optimizing energy consumption.

In the industrial sector, AI-driven IoT solutions are pioneering the realm of predictive maintenance. Traditional maintenance schedules often lead to excessive downtime and unnecessary costs. However, AI can analyze data from sensors in machinery to predict when a component is likely to fail. This predictive capability allows companies to perform maintenance just in time, mitigating potential failures and thus enhancing operational efficiency. The implications of this are vast, extending to reduced costs, improved safety, and longer equipment lifespan.

Another domain where AI significantly boosts IoT is healthcare. IoT devices like smartwatches and fitness trackers generate a wide range of health data. When AI enters the equation, it enables personalized health insights and predictive analytics. For instance, AI algorithms can detect irregular heartbeats from smartwatch data long before a physician might. This early detection can lead to prompt medical intervention, potentially saving lives. Moreover, AI can personalize fitness and diet recommendations based on the continuous analysis of your activity levels and biometrics.

AI's enhancement of IoT also plays a pivotal role in smart cities. Urban areas around the globe are leveraging AIoT to develop sustainable, efficient, and livable environments. AI processes real-time data from a myriad of sensors across the city to improve traffic management, reduce energy consumption, and enhance public safety. Intelligent traffic systems, guided by AI, analyze traffic flows and adjust traffic signals dynamically, minimizing congestion and reducing carbon emissions. On the safety front, AI-powered cameras and sensors can detect unusual activities or hazards, triggering alerts to authorities for quick response.

For environmental monitoring, AI integrated with IoT systems offers significant advancements. Sensors deployed in various environments collect data on air quality, water levels, and weather conditions. AI processes this data to predict environmental changes and potential natural disasters. These predictive abilities are invaluable for early warning systems, enabling proactive measures to minimize damage and enhance community resilience. AI's capability to turn vast and complex data sets into actionable insights is transforming environmental monitoring into a more proactive rather than reactive exercise.

In retail, AI enhances IoT by enabling a more personalized and efficient customer experience. IoT devices in stores collect data on customer behavior, preferences, and shopping patterns. AI analyzes this data to provide personalized product recommendations, optimize inventory management, and streamline supply chain operations. For instance, smart shelves equipped with sensors can alert management when stock levels are low, and AI can predict future inventory needs based on historical sales data and current trends, ensuring shelves are always stocked with in-demand products.

The agricultural sector is also witnessing remarkable changes due to the integration of AI and IoT. Precision farming, enabled by AIoT, allows for the meticulous monitoring and management of crops and soil health. Sensors in the field collect data on soil moisture, temperature, and crop growth, while AI algorithms analyze this data to provide real-time insights and recommendations. Farmers can then make informed decisions about irrigation, fertilization, and pest control, leading to higher yields and more sustainable farming practices.

AI's role in enhancing IoT extends to enhancing cybersecurity. As IoT devices proliferate, they become attractive targets for cyberattacks. AI-driven algorithms can monitor network traffic patterns and device behaviors, quickly identifying anomalies that may indicate a security threat. This proactive approach to cybersecurity ensures that threats are detected and mitigated before they can cause significant harm. By continuously learning from new threats and adapting security protocols, AI ensures that IoT ecosystems remain secure.

Moreover, AI enhances IoT through the concept of edge computing. Traditional IoT systems often rely on sending data to centralized cloud servers for processing, which can introduce latency and bandwidth issues. AI at the edge processes data locally on the device or close

Charlie Morgan

to it, enabling real-time responses and reducing the need for constant cloud communication. This is particularly crucial for applications where latency can impact performance, such as autonomous vehicles and industrial automation.

In the transportation industry, AI-enhanced IoT is revolutionizing everything from fleet management to autonomous driving. AI processes data from vehicle sensors, traffic signals, and GPS to optimize routes, reduce fuel consumption, and improve safety. In fleet management, AI algorithms analyze vehicle performance data to predict maintenance needs and optimize fleet operations. For autonomous driving, AI processes sensor data in real-time to navigate complex environments, avoid obstacles, and ensure passenger safety.

Furthermore, the benefits of AI-enhanced IoT extend into resource management. Smart grids, powered by AIoT, are transforming the energy sector by optimizing energy distribution and consumption. AI algorithms analyze data from smart meters, sensors, and weather forecasts to balance energy supply and demand, reduce energy waste, and integrate renewable energy sources more effectively. This leads to more efficient and sustainable energy systems that benefit both consumers and the environment.

In financial services, AI-powered IoT devices are transforming customer interactions and transaction processing. Smart ATMs and banking kiosks equipped with IoT sensors and AI can provide personalized services, detect fraudulent activities, and streamline operations. AI-driven chatbots analyze customer queries from IoT-enabled devices to provide accurate and quick responses, enhancing customer satisfaction and operational efficiency.

In education, AI and IoT are creating smart learning environments. IoT devices in classrooms collect data on student engagement and performance, while AI analyzes this data to provide personalized learning experiences. Smart classroom equipment, powered by AIoT,

24

can adjust lighting, temperature, and teaching materials based on student needs, creating optimal learning conditions. AI-driven analytics also help educators identify students who may need additional support, enabling timely and targeted interventions.

Entertainment and media industries are also seeing transformative changes due to AI-enhanced IoT. Smart home entertainment systems, powered by AI, learn user preferences to

CHAPTER 3:
SMART HOMES AND AI INTEGRATION

In this brave new world of interconnected devices, smart homes stand at the frontier of AI integration, redefining convenience and efficiency within the household. By seamlessly merging AI with IoT, home automation systems are now capable of learning our routines, optimizing energy consumption, and even enhancing security. Imagine waking up to a home that adjusts the thermostat to your preferred morning temperature, brews coffee, and starts your favorite playlist—all orchestrated by AI-driven algorithms. Security systems imbued with artificial intelligence offer real-time surveillance, facial recognition, and predictive analytics to preempt potential breaches, making your home not just smarter but safer too. Energy management is another critical aspect; AI can analyze usage patterns to reduce waste, lower electricity bills, and contribute to a greener planet. The capabilities of AI in smart homes are evolving rapidly, promising to make daily living not just easier, but smarter at almost every level.

Home Automation Systems

Home automation systems have become the cornerstone of modern smart homes, transforming the way we live and interact with our personal spaces. Powered by AI and connected by IoT, these systems are designed to enhance convenience, efficiency, and safety. They seamlessly integrate a variety of gadgets and appliances, enabling homeowners to control everything from lighting and climate to security and entertainment with a simple voice command or a tap on their smartphone.

One of the most captivating aspects of home automation is its capability to learn from user behavior. Utilizing advanced machine learning algorithms, home automation systems can recognize patterns in daily routines. For instance, your smart thermostat might learn that you prefer a cooler temperature in the evening and adjust itself automatically. Over time, these systems become more intuitive, providing a truly personalized living experience.

Imagine walking into a room where the lights dim to your preferred setting, the blinds adjust to let in just the right amount of natural light, and your favorite playlist starts in the background. This isn't a scene from a science fiction movie; it's the everyday reality made possible by integrated home automation systems. These systems act as the central nervous system of your home, coordinating multiple devices to work in harmony.

The scope of home automation extends to energy management as well. Smart homes are equipped with energy-efficient devices that can be programmed to operate during off-peak hours, reducing electricity bills and minimizing environmental impact. Homeowners can monitor their energy consumption in real-time through apps, making it easier to adopt sustainable practices. For example, intelligent water heaters can heat water only when needed, and smart refrigerators can adjust their cooling cycles based on the time of day.

Security is another critical area where home automation shines. AI-powered security cameras and smart locks provide unparalleled peace of mind. These systems can detect unusual activities, send instant alerts to homeowners, and even notify authorities if necessary. Smart locks eliminate the need for physical keys and can be controlled remotely, allowing homeowners to grant access to visitors or service providers without being present.

Entertainment has seen a revolution as well, with smart home theaters and multi-room audio systems becoming more accessible. These entertainment systems can be effortlessly controlled by voice commands or through integrated apps, providing a seamless and immersive experience. Imagine being able to dim the lights and start your favorite movie with a simple voice command without lifting a finger.

The kitchen is another space where automation has made significant strides. Smart appliances like refrigerators, ovens, and coffee makers are now commonplace. These gadgets can be controlled remotely, and some even come with built-in AI to suggest recipes based on the ingredients you have. Your refrigerator can notify you when you're running low on essentials, and your oven can preheat itself based on the meal you plan to cook.

Voice assistants like Amazon Alexa, Google Assistant, and Apple's Siri have become integral to home automation systems. These AI-powered assistants can execute a plethora of tasks, such as setting reminders, providing weather updates, and controlling smart home devices. By integrating with various smart home platforms, these assistants serve as the central hub, simplifying the control of multiple home automation systems through voice commands.

The integration of AI and IoT in home automation systems also paves the way for improved accessibility. For individuals with disabilities, smart homes can offer a degree of independence previously unimaginable. Voice-controlled devices, automated lighting, and smart doorbells that provide visual and audio notifications are just some of the features that enhance the quality of life for those with mobility challenges.

It's important to consider the interoperability of these systems as well. The market is flooded with a plethora of devices from different manufacturers, and they all need to communicate flawlessly. Standards like Zigbee and Z-Wave are instrumental in ensuring that various devices can work together harmoniously. Interoperability not only simplifies setup but also enhances the user experience by providing unified control over the entire home automation ecosystem.

Despite the numerous benefits, challenges do exist. The initial cost of setting up a comprehensive home automation system can be high. Additionally, concerns about data privacy and security are paramount. As these systems collect a significant amount of personal data to function effectively, safeguarding this information from unauthorized access is crucial. Robust encryption protocols and regular security updates are essential to mitigating these risks.

As technology continues to advance, the future of home automation looks promising. Concepts like augmented reality (AR) and virtual reality (VR) are poised to revolutionize the way we interact with our homes. Imagine using AR glasses to visualize how new furniture would look in your living room or a VR headset to simulate different lighting scenarios before permanent installation. The possibilities are limitless and exciting.

Moreover, with the advent of 5G, the speed and reliability of connected devices will see substantial improvements. Faster data transmission rates and lower latency will enable even more sophisticated applications, from real-time video analytics for security to instantaneous control of household devices. The synergy between AI, IoT, and 5G will redefine home automation, making it more responsive and efficient.

In conclusion, home automation systems represent a significant leap forward in how we manage and experience our living environments. By combining AI and IoT, these systems provide unmatched convenience, security, and energy efficiency, while also adapting to our unique lifestyles. As technology evolves, the integration of advanced features will further enhance the smart home experience, turning our homes into intelligent, responsive spaces that cater to our every need.

AI-Powered Security Solutions

As we delve deeper into the realm of smart homes, one of the most transformative aspects of AI integration is undoubtedly in security solutions. Imagine a home that not only responds to your commands but also watches over you with a level of vigilance and accuracy that's almost human. AI-powered security solutions are revolutionizing how we perceive and manage safety in our personal spaces, making them more intuitive, responsive, and proactive than ever before.

At the heart of AI-powered security solutions is the ability to process vast amounts of data quickly and efficiently. Traditional security systems rely heavily on pre-set parameters and human intervention. Cameras record footage, alarms sound off when sensors detect unusual activity, and it's up to a human to interpret the context and take action. AI changes this paradigm entirely by incorporating not just data collection, but real-time data analysis and decision-making capabilities.

Consider facial recognition technology. This AI-driven feature allows security systems to differentiate between known residents and unknown visitors. When an unfamiliar face is detected, AI algorithms can notify homeowners via their smartphones while simultaneously recording the event for further review. This kind of precision drastically reduces false alarms, as the system becomes smarter and more attuned to the nuances of daily life within the home.

Then there are the smart locks, enabled by AI, that go beyond basic keyless entry. These intelligent systems can understand usage patterns and automatically lock doors when everyone has left the house. They can also integrate with facial recognition technology to allow access only to known individuals, thus adding an additional layer of security. AI-powered locks can even provide temporary access codes for guests or service workers, ensuring that homeowners maintain control over who enters their homes.

AI is also raising the bar with intrusion detection systems. Traditional motion sensors might trigger alerts for any movement, but AI-enabled systems can distinguish between a burglar and a pet or a tree branch moving in the wind. This capability minimizes unnecessary alarms and ensures that when you do receive an alert, it's much more likely to be legitimate and actionable. With machine learning algorithms, the system continually learns and adapts to the environment, enhancing its accuracy over time.

An exciting frontier in AI-powered security solutions is predictive analysis. Instead of just responding to potential threats, these systems can predict them. By analyzing historical data and detecting patterns, AI can identify unusual activity or potential security breaches before they occur. For example, if an unfamiliar vehicle repeatedly passes by in front of the house, AI algorithms can flag this as suspicious and alert homeowners or even notify local authorities if necessary.

Moreover, speech recognition and natural language processing allow AI systems to understand and execute vocal commands, making interaction seamless. Imagine commanding your security system through your voice to lock doors, turn on surveillance cameras, or alert emergency services. AI can also monitor voice patterns and detect stress or duress, offering another layer of security in critical situations.

AI's integration into home security isn't confined to new homes or tech-savvy individuals alone. Retrofit solutions are available, allowing homeowners to upgrade their existing security systems with AI capabilities. For instance, an older CCTV system can be enhanced with AI software to enable real-time facial recognition and behavioral analysis. This flexibility makes advanced security accessible to a wider audience.

Connectivity between different smart devices amplifies the security effectiveness of a smart home. When AI is integrated across multiple systems, such as lighting, heating, and door sensors, these can work in unison to create a comprehensive security network. For instance, if a potential intruder is detected, AI can activate outdoor lights, lock doors, and send alerts simultaneously. This coordinated response can be far more effective than isolated systems working independently.

The integration of AI in security doesn't just stop at hardware; it's also intensely tied to software solutions. Applications on smartphones and tablets allow homeowners to monitor and control their home security systems remotely. These apps utilize machine learning to provide insights and recommendations, such as optimal times for system checks or notifications when it might be time to replace batteries in your smart locks or sensors.

Yet, with all this technological advancement, it's crucial to address privacy concerns. The very capabilities that make AI-powered security systems so effective also necessitate stringent measures to protect users' data. Companies developing these technologies must prioritize data encryption and ensure that sensitive information remains secure from potential cyber threats. Implementing robust security protocols and being transparent about data usage can help in building consumer trust, ensuring these advanced systems are both effective and secure.

Additionally, collaboration between various stakeholders, including tech companies, security firms, and regulatory bodies, is essential for developing standardized protocols. These standards can ensure compatibility across different devices and systems, providing consumers with a smoother experience. They're also vital for addressing ethical considerations, ensuring that AI-powered security solutions are fair, transparent, and respectful of individual privacy.

In conclusion, AI-powered security solutions represent a significant leap forward in the quest to secure our homes. They offer unparalleled accuracy, predictive capabilities, and a level of convenience that traditional systems cannot match. As AI technology continues to advance, we can expect these systems to become even more integral to our daily lives, providing a reassuring blend of safety, efficiency, and peace of mind in our smart homes.

Energy Management

When we think about the modern smart home, energy management is one of its most critical aspects. Incorporating AI and IoT for energy efficiency not only makes homes more sustainable but also significantly reduces utility bills. At its core, energy management in smart homes involves the monitoring, control, and optimization of energy usage across various devices and systems. With AI algorithms learning and predicting usage patterns, and IoT devices executing precise control mechanisms, energy management has never been more intelligent or efficient.

One of the key applications of AI in energy management is in predictive analytics. AI systems can analyze historical data to anticipate energy consumption patterns. For instance, during the summer months, the demand for air conditioning might spike. The AI can predict this surge and optimize the cooling schedule to ensure that the house is pre-cooled during less expensive off-peak hours. This predictive ability contributes to a more balanced and efficient use of electricity, which is crucial for both savings and reducing the strain on the grid.

On a smaller scale, smart thermostats are instrumental in energy management. Unlike traditional thermostats, smart thermostats use algorithms to learn the preferences and routines of the home's inhabitants. Over time, the thermostat can automatically adjust the temperature to suit occupancy patterns, reducing unnecessary heating or cooling when no one is home. Such devices can be controlled remotely via smartphones, adding an additional layer of convenience and control.

But energy management isn't limited to heating and cooling systems. AI and IoT also extend to lighting. Smart lighting systems can determine when rooms are occupied, adjusting the brightness or turning off the lights entirely when a room is vacant. Some systems can even adjust the lighting based on the time of day or ambient light levels, providing optimal illumination without wasting electricity. These small adjustments collectively make a significant impact on overall energy consumption.

Additionally, integrating renewable energy sources into smart homes enhances energy management. Solar panels, for instance, can be monitored and controlled through a central AI-based management system. The system can determine the best times for harvesting solar energy and decide when to store it in batteries or feed it back into the grid. By managing energy flow more intelligently, homeowners can maximize their use of renewable sources and minimize reliance on traditional power grids.

Energy storage is another vital component. As battery technology improves, more homeowners are incorporating energy storage solutions into their smart homes. AI can play a crucial role in managing these batteries, determining the best times to charge or discharge based on usage patterns and electricity rates. For example, an AI system might decide to store excess energy generated by solar panels during the day and use it during peak hours when electricity rates are higher, offering both economic and environmental benefits.

Moreover, smart appliances are at the forefront of energy efficiency. Modern refrigerators, washing machines, and even ovens are becoming increasingly intelligent. These devices can operate during off-peak hours to reduce energy costs. Washing machines, for instance, can delay the start of a wash cycle until electricity demand and rates are lower. Meanwhile, smart refrigerators can optimize their cooling cycles to maintain freshness while consuming minimal energy.

For homeowners keen on tracking their energy usage, IoT-enabled energy monitors provide real-time insights. These devices offer detailed reports on energy consumption patterns, highlighting opportunities for improvement. By understanding which devices or systems are costing the most, homeowners can make informed decisions about upgrades or behavioral changes to enhance efficiency. Some energy monitors also incorporate machine learning algorithms to provide personalized recommendations for reducing energy usage.

Utility companies are also leveraging AI and IoT for demand response programs. By partnering with smart home systems, utilities can dynamically manage grid loads, reducing the likelihood of blackouts and enhancing overall grid stability. Homeowners can participate in these programs, receiving financial incentives for allowing temporary adjustments to their energy usage during peak demand periods. AI-driven systems ensure that any adjustments minimize discomfort and inconvenience for residents.

In the realm of water heating, smart water heaters are taking center stage. These appliances can be programmed to heat water based on usage patterns, ensuring hot water availability when needed while conserving energy during off-peak times. With AI optimization, water heating becomes more responsive and aligned with the household's requirements, reducing wasted energy.

The role of energy management extends beyond individual homes. In smart communities, interconnected homes share energy resources more efficiently. AI systems can manage energy distribution across multiple homes, prioritizing areas with high demand and optimizing the use of shared resources like community solar panels or energy storage facilities. This collective approach amplifies the benefits of smart energy management, making entire neighborhoods more sustainable and resilient.

In conclusion, energy management within smart homes represents a convergence of intelligence, technology, and sustainability. By integrating AI and IoT, homeowners can achieve unprecedented levels of efficiency and control, contributing to both economic savings and environmental stewardship. As technology continues to advance, the possibilities for smarter, more efficient energy management will expand, paving the way for a more sustainable future.

CHAPTER 4:
INTELLIGENT CITIES

As urban landscapes grow increasingly complex, the fusion of AI and IoT brings forth Intelligent Cities that redefine our interaction with the urban environment. By tapping into vast networks of interconnected sensors and leveraging advanced algorithms, these cities promise enhanced efficiency in governance, unparalleled public safety, and robust infrastructure management. Smart transportation systems optimize traffic flow and reduce commute times, while real-time data collection helps in proactive maintenance of public utilities. Meanwhile, intelligent urban planning facilitates sustainable development, making cities not just smarter, but also greener and more livable. Surveillance and public safety measures employing AI ensure quicker response times and more efficient law enforcement, creating a safer environment for inhabitants. Intelligent Cities represent a harmonious blend of technology and human experience, paving the way for a future where urban living is synonymous with innovation, sustainability, and enhanced quality of life.

Urban Planning and AI

Urban planning is the keystone of modern cities, orchestrating the balance between infrastructure, resources, and the needs of residents. The integration of Artificial Intelligence (AI) into urban planning is transforming how cities are designed, managed, and evolved. With the accelerated pace of urbanization, cities face complex challenges ranging from overcrowding and pollution to resource management and public safety. AI offers unprecedented opportunities to address these challenges through data-driven and intelligent solutions.

Historically, urban planning relied heavily on static models and manual data collection. Today, AI-driven tools enable urban planners to access real-time data and predictive analytics. This shift has led to more dynamic decision-making processes. Urban planners now use AI to analyze vast datasets from various sources like satellite imagery, traffic sensors, and social media feeds. This analysis provides insights that inform the planning and development of more efficient, sustainable, and livable cities.

One of the most significant impacts of AI on urban planning is in the realm of traffic management. Traditional traffic systems struggle to cope with the growing number of vehicles on the road. AI-powered traffic management systems can analyze traffic patterns in real-time, predict congestion, and implement adaptive traffic control strategies. These systems optimize traffic flow, reducing commute times and lowering emissions. For example, AI algorithms can adjust traffic signal timings based on real-time traffic conditions, making intersections more efficient and improving overall traffic flow.

Beyond traffic management, AI is instrumental in environmental monitoring and sustainability efforts within cities. Urban areas are often hotspots for pollution and resource consumption. AI systems can monitor air and water quality, track waste management processes, and optimize energy usage. By predicting pollution levels, AI can assist in devising strategies to mitigate environmental impacts, such as enforcing emissions regulations or improving public transportation options.

Another crucial area where AI aids urban planning is resource allocation. Cities must efficiently distribute resources like water, electricity, and public services to sustain their populations. AI-driven models can predict future demand for these resources and optimize their distribution. This is particularly important in cities with rapidly growing populations, where the balance between supply and demand is delicate. For instance, AI can help utilities forecast energy consumption patterns and adjust production accordingly, ensuring residents have uninterrupted access to electricity.

AI also plays a pivotal role in urban safety and emergency response. With the proliferation of IoT devices, cities can now deploy AI-driven surveillance systems that monitor public spaces for unusual and potentially dangerous activities. These systems identify patterns that human operators might miss, sending alerts to law enforcement or emergency services as needed. Additionally, AI can assist in disaster management by predicting natural disasters, such as floods or earthquakes, and planning efficient evacuation routes.

Public transportation is another beneficiary of AI in urban planning. Efficient and reliable public transit systems are vital for sustainable urban growth. AI optimizes public transportation routes and schedules based on passenger demand and traffic conditions. This not only enhances service reliability but also encourages more residents to use public transit, reducing the reliance on personal vehicles. Some cities are even experimenting with AI-operated autonomous buses, potentially revolutionizing public transport systems.

Urban housing and real estate development is another domain where AI's influence is increasingly felt. AI algorithms can analyze demographic trends, economic indicators, and housing market data to forecast future housing needs and identify optimal locations for new developments. This helps city planners ensure that residential areas are adequately developed to meet the evolving needs of the population.

AI also facilitates better community engagement in urban planning. Traditionally, residents had limited input into planning decisions. AI-driven platforms can aggregate feedback from social media, surveys, and public forums, providing planners with a deeper understanding of residents' needs and preferences. This leads to more inclusive and democratic urban development processes.

Furthermore, AI's role in economic development within urban areas cannot be understated. By analyzing economic trends and labor market data, AI helps cities create robust economic development plans that attract businesses and create jobs. AI-driven economic models can simulate the impact of different policy decisions, aiding policymakers in making informed choices that foster economic growth.

Intelligent waste management systems powered by AI are transforming how cities handle and reduce waste. Traditional waste management methods are inefficient and often lead to environmental degradation. AI can predict waste generation patterns, optimize collection routes, and identify opportunities for recycling and composting. This leads to more efficient waste disposal processes and a significant reduction in urban pollution.

Public health is yet another area where AI is making strides in urban planning. AI systems can identify and predict public health trends, enabling cities to allocate healthcare resources more effectively. For instance, during disease outbreaks, AI can help track the spread of infections and identify hotspots, enabling rapid response and containment measures. This proactive approach to public health management improves overall community well-being.

In conclusion, AI is revolutionizing urban planning by making cities smarter, more efficient, and more sustainable. As urban areas continue to expand, the integration of AI into urban planning processes is not just advantageous but necessary. It empowers urban planners to make data-driven decisions, optimize resources, enhance public safety, and foster community engagement. By leveraging AI, cities can better anticipate future challenges, adapt to changing needs, and ensure a higher quality of life for their inhabitants. The future of urban planning is undoubtedly tied to the continued advancement and integration of AI technologies.

Smart Transportation Systems

In the bustling fabric of modern cities, the essence of movement is integral. Smart transportation systems are radically changing the landscape of urban mobility, blending AI and IoT to reinvent how we navigate the cityscape. These cutting-edge systems are designed not just to alleviate traffic congestion but to enhance the overall efficiency and sustainability of urban transportation.

At the core of smart transportation lies the dynamic interplay of real-time data, machine learning algorithms, and connected infrastructure—components that collectively create an adaptive, responsive network. Imagine a city where traffic lights adjust their patterns in real-time based on current traffic flow, reducing idle times and fuel consumption. Picture this alongside intelligent traffic management systems that predict and mitigate congestion through continuous analysis of vast datasets obtained from sensors, cameras, and vehicular feedback.

Artificial Intelligence (AI) brings a layer of predictive power to transportation systems, enabling them to foresee traffic bottlenecks before they occur. For instance, data from previous months and seasonal patterns can be analyzed to predict traffic on a holiday weekend. This allows for preemptive adjustments in traffic signal timings and dynamic rerouting suggestions to drivers via connected vehicles and mobile applications.

Moreover, IoT devices penetrate every aspect of the transportation network, from streetlights equipped with sensors to smart parking meters. By constantly communicating with each other, these devices form a cohesive and responsive ecosystem. Intelligent parking solutions, for example, can guide drivers to available spots via mobile apps, significantly reducing the time spent circling blocks searching for parking. This not only enhances convenience but also cuts down emissions from unnecessary driving.

Public transportation is another beneficiary of intelligent systems. Real-time tracking of buses and trains, shared through mobile apps, can keep passengers informed about delays and estimated arrival times, improving the user experience and reliability perception. Additionally, AI-driven analytics can optimize routes and schedules based on passenger needs and flow patterns, ensuring that resources are allocated efficiently while minimizing waiting times.

The integration of autonomous vehicles into the urban transport fabric represents another leap forward. While fully autonomous transport is still on the horizon, pilot programs and semi-autonomous features are being tested and rolled out in various cities. Smart transportation systems here play a crucial role as they provide the necessary infrastructure and data to facilitate safe and efficient autonomous driving. Vehicle-to-Infrastructure (V2I) communication allows autonomous vehicles to interact with traffic signals, warning systems, and even other cars, leading to smoother and safer traffic conditions.

Beyond personal and public transport, logistics too benefit from smart transportation. AI-enhanced fleet management systems help in optimizing delivery routes, ensuring timely deliveries while cutting operational costs. Predictive maintenance powered by IoT sensors helps in monitoring vehicle health in real-time, reducing downtime and extending the lifespan of fleet vehicles. These systems also enable real-time tracking of goods, enhancing transparency and customer satisfaction.

Environmental impact is another critical aspect addressed by smart transportation systems. By ensuring optimized routes and reducing idling times, these systems contribute significantly to reducing greenhouse gas emissions. Electric and hybrid vehicles are becoming more integrated into intelligent infrastructures, with smart grids facilitating efficient charging and load balancing. These measures collectively cultivate a more sustainable urban environment, aligning with global efforts to combat climate change.

The benefits of smart transportation extend beyond obvious efficiencies and environmental impact. They also encapsulate economic growth. Efficient transportation systems can reduce commute times, allowing people to be more productive. Job creation in technology, data analysis, and infrastructure development sectors are some of the direct economic benefits arising from the deployment of smart transportation solutions.

However, the deployment of these systems isn't devoid of challenges. Data privacy and security remain significant concerns as transportation systems handle a vast amount of sensitive information. Ensuring robust cybersecurity measures while adhering to privacy regulations is essential for gaining public trust and safeguarding against potential attacks. Additionally, the need for interoperability between different systems and standards requires careful planning and a high degree of collaboration among multiple stakeholders.

Another hurdle is the initial cost and complexity of implementing smart transportation systems. While the long-term benefits and savings are evident, the initial investment can be substantial. Cities need to strategize funding, possibly leveraging public-private partnerships, to deploy these advanced systems. Education and training of the workforce are equally crucial components, ensuring that personnel can effectively manage and maintain these sophisticated systems.

Looking ahead, the evolution of smart transportation systems will continue to be driven by technological advancements and growing urban demands. Emerging technologies such as 5G will provide faster and more reliable communication between devices, enhancing the responsiveness and efficiency of smart transportation networks. As AI becomes more sophisticated, its ability to handle complex datasets and provide actionable insights will only improve.

In an ever-evolving urban landscape, smart transportation systems stand as a testament to human ingenuity and the bright future AI and IoT promise. They're not just about moving people or goods from one place to another; they're about creating smarter, more sustainable, and livable cities. With a collaborative, innovative approach, the possibilities are boundless, transforming our everyday commutes into seamless, efficient experiences.

Public Safety and Surveillance

In the realm of intelligent cities, ensuring public safety and surveillance stands as a critical pillar. With the integration of AI and IoT, our urban environments are transforming into responsive, predictive ecosystems capable of enhancing security and public welfare in ways previously unimaginable. The potential for these technologies to revolutionize public safety is immense, offering capabilities that enhance monitoring, emergency response, and crime prevention.

The first layer of this transformation is the deployment of IoT-based surveillance systems. These systems employ a network of interconnected devices such as cameras, sensors, and other data-collecting instruments strategically embedded throughout the city. These devices continuously gather real-time data, monitoring different aspects of the urban environment.

AI algorithms analyze the data collected by these IoT devices. Machine learning models can detect unusual patterns or activities by parsing through vast amounts of footage and sensor data, efficiently identifying potential threats or anomalies. For instance, an AI system in a public park might notice a crowd behaving unusually or unattended packages, prompting immediate alerts to security personnel.

Beyond mere detection, predictive analytics also play a significant role. By analyzing historical data and current trends, AI systems can forecast where and when crimes are likely to happen, allowing law enforcement to deploy resources more effectively. This proactive approach can drastically reduce response times and potentially deter criminal activities before they occur.

Additionally, AI and IoT can enhance emergency response strategies. During natural disasters or severe incidents, real-time data from IoT sensors can provide comprehensive situational awareness. Emergency services can quickly assess the extent of the problem, identify areas most in need, and allocate resources precisely where they're required. Drones equipped with cameras and sensors can further support by surveying affected areas, providing real-time visuals, and aiding in coordination efforts without exposing human responders to danger.

Public transportation hubs also benefit from heightened security. AI-driven surveillance systems monitor transit stations for suspicious behavior or unattended baggage. These systems are designed to provide early warnings, which can trigger escalated security measures or evacuations if necessary. Facial recognition technology, bolstered by AI, can identify individuals flagged in criminal databases, adding an additional layer of protection against potential threats.

However, as we advance public safety through AI and IoT, privacy concerns inevitably arise. The constant monitoring and data collection inherent in such systems can lead to worries about misuse or overreach. Ensuring robust data protection measures and transparent policies is paramount in addressing public concerns and maintaining trust. Strong regulations and ethical guidelines must be put in place to balance safety improvements with individual privacy rights.

Facial recognition and automated number plate recognition (ANPR) systems epitomize the dual-edged nature of surveillance. While these technologies augment law enforcement capabilities, they also introduce significant privacy dilemmas. Proper regulations and oversight must govern their use to prevent misuse and ensure they serve the public interest.

Surveillance isn't restricted to crime prevention; it plays a crucial role in managing public health. During the COVID-19 pandemic, IoT devices and AI analytics were instrumental in monitoring and enforcing social distancing. Thermal cameras and wearable health devices provided data for tracking potential outbreaks and implementing timely interventions, showcasing how integrated technology can safeguard public health on a large scale.

Furthermore, these technologies contribute to environmental safety. Air quality sensors distributed throughout cities can detect harmful pollution levels and immediately notify authorities to take corrective action. Noise monitoring systems similarly help manage urban noise pollution, enhancing the overall quality of life for residents.

Another realm where public safety and AI converge is traffic management. Intelligent traffic systems use real-time data from cameras, motion sensors, and connected vehicles to optimize traffic flow and reduce accidents. These systems can dynamically adjust traffic signals, provide real-time updates to commuters, and coordinate with emergency vehicles to ensure safe and efficient travel through busy urban landscapes.

Enhancing public safety through AI and IoT isn't without challenges. Technical issues such as integration complexity, data interoperability, and system scalability present significant hurdles. Cities must navigate these challenges to build cohesive, effective surveillance networks. Successful implementations often require collaboration between multiple stakeholders, including government agencies, private companies, and community groups.

Public awareness and education are also crucial. Citizens should understand how these technologies function and the benefits they present. Engagement and transparency can foster public trust and encourage community participation in safety initiatives. When people are informed, they're more likely to support and collaborate with efforts to enhance urban security.

Cities around the world are already reaping the benefits of these advancements. In Singapore, the government has implemented the Smart Nation initiative, using a network of interconnected sensors and cameras to monitor various city aspects, from water levels and energy consumption to traffic and public safety. Similarly, New York uses an array of IoT devices and AI-driven analytics to improve emergency response, reduce crime, and manage massive public events effectively.

In summary, public safety and surveillance in intelligent cities are undergoing a profound transformation thanks to the synergistic power of AI and IoT. The ability to collect, analyze, and act on vast amounts of real-time data is revolutionizing how urban centers prevent crimes, respond to emergencies, and safeguard public health. While challenges such as privacy concerns and technical complexities remain, the potential benefits to safety and quality of life are substantial. Ensuring the ethical application of these technologies will be key to building smarter, safer cities that inspire confidence and foster a sense of security for all residents.

CHAPTER 5:
HEALTHCARE REVOLUTION

The integration of AI and IoT is transforming healthcare into a more precise, efficient, and personalized field, marking a revolutionary shift in how medical services are delivered and experienced. AI-driven diagnostics and imaging are enabling faster, more accurate detections of ailments, often identifying issues long before they become critical. Simultaneously, IoT in patient monitoring is providing real-time data that doctors can analyze to make well-informed decisions, minimizing the risk of human error. Personalized medicine, empowered by wearables, is tailoring treatments to individual genetic profiles and lifestyles, thus maximizing efficacy and reducing side effects. These advancements collectively signify a monumental leap forward in patient care, making healthcare not just reactive but proactive and predictive as well.

AI in Diagnostics and Imaging

In recent years, AI has become an integral part of diagnostic and imaging processes, reshaping the healthcare industry in ways that were once deemed impossible. Traditional diagnostic techniques and imaging methods have always been fundamental in guiding effective treatment plans. However, the introduction of AI has exponentially increased the accuracy, speed, and efficiency of these processes, providing a clear demonstration of how technology can spearhead a healthcare revolution.

The application of AI in diagnostics and imaging primarily involves sophisticated algorithms that can analyze medical data and recognize patterns far beyond human capability. Machine learning models are trained on vast datasets of medical images, enabling them to identify minute details that may be easily overlooked by the human eye. This meticulous level of detail ensures that diagnoses are not only more accurate but also quicker, which is crucial in time-sensitive medical conditions like stroke or cancer.

One of the most influential advancements has been the use of convolutional neural networks (CNNs) in medical imaging. CNNs have proven effective in analyzing and interpreting complex visual data, and when applied to medical imaging, they can detect anomalies such as tumors or lesions with a high degree of precision. For instance, in radiology, AI-driven tools can flag potential areas of concern in mammograms or CT scans, allowing radiologists to focus their expertise on confirming these findings and determining the best course of action.

The benefits of integrating AI into diagnostic imaging extend to various medical fields. Ophthalmology has seen significant improvements with AI algorithms that can screen for diabetic retinopathy by analyzing retinal images. Early detection is crucial in preventing vision loss, and AI provides a faster, more reliable means of screening large populations. Similarly, in dermatology, AI tools can evaluate skin lesions and moles, differentiating between benign and malignant conditions with impressive accuracy. This capability is especially useful in remote areas where access to specialists might be limited.

AI's role in diagnostics is not just limited to image analysis. Natural language processing (NLP) technologies are being leveraged to sift through electronic health records (EHRs), extracting relevant patient data to provide comprehensive case reviews. This can help in generating more personalized and accurate diagnostic reports by integrating historical patient information with current imaging data.

The integration of AI in diagnostics and imaging also paves the way for predictive analytics. By utilizing historical data and identifying patterns, AI can forecast disease progression and potential medical outcomes. This enables healthcare providers to adopt a more proactive approach, focusing on prevention and early intervention rather than reactive treatment. For example, AI models can evaluate a patient's risk of developing certain conditions based on genetic, lifestyle, and environmental factors, allowing for tailored preventive measures or early treatments to be employed.

Despite these advancements, the implementation of AI in diagnostics and imaging is not without challenges. One significant hurdle is the need for vast, high-quality datasets to train AI models. These datasets must be representative of diverse populations to ensure that AI tools are effective across different demographics. Additionally, the integration of AI systems with existing medical infrastructure can be complex and costly, requiring significant investment and training for medical staff to effectively use these new technologies.

Moreover, regulatory considerations must be addressed. AI tools used in diagnostics and imaging must comply with stringent medical standards and obtain approval from relevant regulatory bodies to ensure their safety and efficacy. There is also the matter of ethical considerations, ensuring patient privacy and data security as these systems process sensitive health information.

Nonetheless, the potential benefits of AI in diagnostics and imaging far outweigh these challenges. AI-driven tools can significantly reduce the burden on healthcare systems by automating routine tasks, allowing medical professionals to focus on more complex patient care. This not only increases efficiency but also improves patient outcomes by enabling faster and more accurate diagnoses.

Furthermore, the synergy between AI and IoT is unlocking even greater possibilities in diagnostics and imaging. IoT devices can continuously monitor patients' health metrics and feed this data into AI systems for real-time analysis. This ongoing data collection and monitoring can lead to more timely diagnoses and personalized treatment plans, enhancing the overall quality of care.

The future of diagnostics and imaging lies in the continual evolution and integration of AI technologies. Emerging innovations such as quantum computing and advanced machine learning techniques promise even greater capabilities in processing and analyzing medical data. Ultimately, the goal is to create healthcare environments where AI assists in ensuring every patient receives accurate, timely, and individualized care, revolutionizing our approach to medical diagnostics and imaging.

This intersection of AI and healthcare is not just a technological leap; it's a paradigm shift that emphasizes the importance of innovation in improving human health. The ongoing development and application of AI in diagnostics and imaging illustrate the profound impact that technology can have on healthcare, setting the stage for a more efficient and effective medical practice for generations to come.

IoT in Patient Monitoring

In the rapidly evolving landscape of healthcare, the integration of the Internet of Things (IoT) in patient monitoring is proving to be a game-changer. Despite the complexities, the resulting benefits in terms of patient care, early diagnosis, and efficient management of chronic diseases are indisputable. At the core of this transformation is the ability to collect real-time data continuously and reliably, offering insights that were previously incomprehensible.

The transformation begins at the patient's bedside. Traditional methods of patient monitoring often involve manual observation and periodic measurements. IoT devices, on the other hand, capture continuous streams of data such as heart rate, oxygen levels, blood pressure, and glucose levels. These devices range from wearable sensors to implantable devices, creating a network of interconnected health monitors. The real-time data captured can be immediately analyzed and sent to healthcare providers, allowing for quick intervention if necessary.

This kind of continuous monitoring is particularly beneficial for managing chronic conditions. For example, patients with diabetes can use IoT-enabled glucose monitors that provide continuous feedback on their blood sugar levels. This helps in the fine-tuning of medication and lifestyle adjustments on a day-to-day basis, potentially averting severe complications. Likewise, cardiac patients benefit greatly from wearable ECG monitors that can detect irregular heartbeats well before they become life-threatening issues.

On the technological front, sensor accuracy and data transmission reliability are crucial. High-quality sensors are designed to be minimally invasive and capable of robust data collection in various environmental conditions. These sensors transmit data via secure networks, ensuring privacy and integrity. Advancements in Bluetooth Low Energy (BLE) and Zigbee protocols help in reducing power consumption while extending battery life, making these devices practical for long-term use.

Artificial Intelligence (AI) further amplifies the capabilities of IoT in patient monitoring. AI algorithms can analyze the colossal streams of data generated by IoT devices to identify patterns and anomalies that might escape human eyes. These algorithms are particularly adept at predictive analysis, which means potential health issues can be identified and addressed before they escalate. This predictive capability is transforming reactive healthcare into proactive health management.

Consider a real-world scenario where an elderly patient's vital signs are continuously monitored through an array of IoT devices. Any deviation from the norm, such as a sudden spike in heart rate or a drop in oxygen levels, triggers an alert. This alert can prompt healthcare providers to take immediate action, such as sending an ambulance to the patient's location. Such responsiveness can be vital for reducing emergency response times and improving the chances of positive outcomes.

The data collected by IoT devices also enhances personalized medicine. By analyzing the unique health data of an individual over time, healthcare providers can tailor treatments and medication regimens that are highly specific to the patient's needs. This kind of personalized approach not only improves effectiveness but also minimizes side effects and complications.

However, the implementation of IoT in patient monitoring does come with its fair share of challenges. Data security and patient privacy are paramount concerns. Robust encryption methods and stringent data protection regulations are essential to ensure that sensitive health data is safeguarded. There is also the challenge of interoperability—various devices from different manufacturers need to work seamlessly together and integrate into existing healthcare systems.

Technological advancements and standardization efforts are underway to address these challenges. Open-source platforms and standardized communication protocols are facilitating better integration and data sharing. Moreover, collaborations between tech companies and healthcare providers are helping to create an ecosystem where IoT in patient monitoring can flourish.

The ethical implications of continuous patient monitoring also require careful consideration. While the benefits are clear, there is a fine line between helpful monitoring and intrusive surveillance. Patients should have control over who can access their data and how it's used. Transparent policies and informed patient consent are crucial in navigating these ethical waters.

Despite these challenges, the potential for IoT in patient monitoring to revolutionize healthcare is immense. Remote monitoring, for example, can significantly reduce the need for hospital visits, thereby lowering healthcare costs and freeing up medical resources. This is especially valuable in rural areas where access to healthcare facilities might be limited.

Furthermore, the aggregation and analysis of data from numerous patients can provide researchers with valuable insights into the progression of diseases and the effectiveness of treatments on a population level. This could expedite the development of new drugs and therapies.

The future of IoT in patient monitoring is bright, with continuous advancements promising even more sophisticated monitoring solutions. Implantable sensors that can monitor biochemical changes at the cellular level, for instance, are on the horizon. The integration of AI becomes more refined with each iteration, pushing the boundaries of what's possible in predictive and personalized healthcare.

In conclusion, the incorporation of IoT in patient monitoring is not just an incremental improvement but a monumental leap in how healthcare is delivered and managed. By harnessing real-time data, predictive analytics, and personalized care strategies, we are poised to improve health outcomes on an unprecedented scale. The journey is fraught with challenges, but the destination is a future where healthcare is more responsive, preventive, and personalized.

Personalized Medicine and Wearables

In today's rapidly evolving healthcare landscape, personalized medicine and wearable technology are not just catching the spotlight; they're entirely transforming it. The integration of AI and IoT in these domains offers unprecedented levels of customization and intelligence, enabling healthcare that's tailored to each individual. Imagine bypassing traditional "one-size-fits-all" medical strategies for treatments and interventions crafted specifically for your genetic makeup, lifestyle, and long-term health data. This is the promise of personalized medicine paired with the power of wearables.

Personalized medicine is fundamentally about understanding that every individual is unique, not just in terms of genetics but also in their lifestyle, environment, and health history. With advancements in genomics, big data analytics, and artificial intelligence, we now have the tools to analyze vast amounts of health-related data. This process enables us to deliver more precise, predictive, and preventive healthcare solutions. The idea is straightforward: by leveraging AI algorithms to analyze a person's unique traits, medical professionals can predict more accurately which treatments will be most effective.

Wearable devices play a crucial role in this transformation. These gadgets, ranging from smartwatches and fitness trackers to more advanced medical sensors, continuously monitor and collect health data in real-time. It's not just about counting steps or measuring heart rates anymore. Modern wearables can track a wide array of biometrics such as blood glucose levels, oxygen saturation, and even electrocardiogram (ECG) signals. The data gathered from these devices can be transmitted to cloud-based systems where AI algorithms analyze them to provide insightful recommendations, detect anomalies, and even alert healthcare providers in case of emergencies.

Consider the management of chronic diseases like diabetes or hypertension. Wearable devices equipped with AI capabilities can monitor critical metrics continuously, sending alerts for abnormal readings that might indicate a potential issue. Healthcare providers can use these insights to adjust treatment plans in real-time, ensuring that patients receive the right medication at the right time. This dynamic and adaptive approach reduces the risk of complications and enhances the overall quality of life for patients.

Moreover, personalized medicine takes advantage of genetic data to offer custom-tailored therapies. Advances in genetic sequencing technology have made it affordable and accessible to decode individual genomes. AI and machine learning algorithms then analyze this genomic data to identify predispositions to specific conditions or likely responses to certain treatments. For example, in oncology, AI-driven algorithms can help analyze the genetic mutations within cancer cells to suggest the most effective treatment regimens. This level of specificity improves treatment outcomes and minimizes side effects.

A practical example of the synergy between AI, IoT, and personalized medicine can be seen in cardiac care. Traditional methods of diagnosing heart conditions often rely on intermittent visits to a cardiologist and sporadic snapshots of heart health. Contrast this with a scenario where a patient wears a smart device equipped with AI that continuously monitors heart activity. This device can detect early signs of arrhythmia or other anomalies and alert both the patient and their healthcare provider long before a crisis point. This ongoing monitoring enables more proactive and preemptive care interventions, potentially saving lives.

Psychiatric conditions also benefit from this technological confluence. Wearables can monitor physiological indicators like sleep patterns, physical activity, and circadian rhythms, which are often closely tied to mental health. AI algorithms can cross-reference this data with reported mood changes, medication adherence, and lifestyle factors to provide a comprehensive mental health overview. Enhanced by these insights, mental health professionals can tailor their therapeutic strategies uniquely suited to each patient, enhancing treatment efficacy and personal well-being.

Privacy and ethical considerations cannot be overlooked when discussing personalized medicine and wearables. The vast amounts of sensitive data generated necessitate robust cybersecurity measures to protect patient information. Ensuring data accuracy and preventing misuse is paramount. AI ethics committees and regulatory frameworks play crucial roles in developing standards that balance innovation with patient rights. Transparent policies and proactive measures can build trust, ensuring these technologies' widespread and beneficial adoption.

Incorporating AI and IoT into personalized medicine also democratizes healthcare, making high-quality, individualized care accessible to a broader population. Wearable devices equipped with AI capabilities can function independently in remote or underserved areas, bringing sophisticated diagnostic and monitoring tools to patients who might otherwise lack regular healthcare access. This advancement is particularly significant for managing pandemic outbreaks or delivering healthcare in resource-limited settings, where conventional healthcare infrastructure might be lacking.

The integration of wearables into personalized healthcare has also fostered a culture of proactive health management. Patients can now track their day-to-day health metrics, providing motivation to maintain healthier lifestyles. This shift from reactive to proactive healthcare not only mitigates chronic conditions but also promotes a sense of agency and engagement in one's health journey. AI-powered wearables offer personalized feedback and actionable insights, encouraging individuals to make healthier choices actively.

Moving forward, the realm of personalized medicine and wearables is set to expand further, fueled by advancements in AI and IoT. Researchers are exploring the potential of integrating additional biosensors into wearable devices, enabling even more comprehensive health monitoring. Moreover, the fusion of virtual reality (VR) and augmented reality (AR) with wearable technology promises to take personalized care to the next level, offering immersive and interactive healthcare experiences.

As personalized medicine becomes more prevalent, the traditional healthcare ecosystem will undergo a significant transformation. The role of healthcare providers will evolve from being primarily reactive to becoming highly proactive, aided by continuous real-time patient data and predictive analytics. Hospitals and clinics might transition from multi-purpose facilities to more specialized centers focusing on acute and intensive care, with routine monitoring and management handled remotely.

The impact of personalized medicine and wearables extends beyond individual health outcomes, influencing public health at large. Aggregated and anonymized data from wearables can provide valuable insights into population health trends, enabling more effective public health strategies and interventions. This data-driven approach can help predict and mitigate the impact of health crises on a large scale, ultimately leading to more resilient healthcare systems.

Nevertheless, achieving the full potential of personalized medicine and wearables will require ongoing collaboration among technologists, healthcare professionals, policymakers, and patients. Interdisciplinary initiatives will foster innovation, while citizen engagement and empowerment will drive demand and acceptance. As these technologies evolve, so too will our understanding and expectations of healthcare, leading to a future where personalized, data-driven medicine is the norm rather than the exception.

In conclusion, the convergence of AI, IoT, and wearables marks the dawn of a new era in healthcare. By empowering individuals with personalized insights and fostering data-driven decision-making, these technologies promise a future where healthcare is not only more effective but also more human. Personalized medicine and wearables exemplify how technology can be harnessed to meet individual needs, enhancing both quality of life and the overall efficacy of healthcare systems.

CHAPTER 6:
INDUSTRIAL IOT (IIOT)

In the heart of modern industry, the integration of AI and IoT—collectively known as Industrial IoT (IIoT)—is nothing short of revolutionary. By weaving AI's intelligence through the fabric of industrial operations, IIoT is transforming traditional manufacturing, maintenance, and supply chain processes. Imagine factories where machines not only talk to each other but also predict and prevent their own failures, or supply chains so tightly monitored by advanced analytics that inefficiencies are eradicated before they even arise. This evolution is not merely about technology; it's reshaping how industries think, operate, and compete on a global scale. With AI-driven automation and data-driven decision-making, IIoT turns the promise of smarter, leaner, and more responsive manufacturing into a reality, driving unprecedented levels of productivity and innovation.

Manufacturing Automation

In the realm of Industrial IoT (IIoT), manufacturing automation stands as a transformative force reshaping the production landscape. The integration of artificial intelligence (AI) and IoT technologies within the manufacturing sector is driving unprecedented levels of efficiency, precision, and scalability. This convergence is not just an evolution; it's a revolution, redefining how factories operate and how products are made.

Let's begin with a fundamental shift brought about by manufacturing automation: the ability to collect and analyze data in real-time. Traditional manufacturing relied heavily on human oversight and periodic inspections, which left room for errors and inefficiencies. With IIoT, sensors embedded in machinery collect vast amounts of data on every aspect of the production process. This data is then processed by AI algorithms to provide actionable insights. Automated systems can instantly detect anomalies, forecast equipment failures, and optimize production schedules. The result? A seamless operation with minimal downtime and maximum output.

One of the most significant advancements in manufacturing automation is the deployment of robotics. Industrial robots, powered by AI, now handle tasks ranging from simple assembly to complex processing. Unlike their human counterparts, these robots can work tirelessly, maintaining high levels of precision and productivity. Moreover, collaborative robots, or "cobots," are revolutionizing assembly lines by working alongside human operators. Cobots are designed to perform repetitive tasks, reducing the physical strain on workers and allowing them to focus on more skilled activities. This symbiotic relationship between man and machine enhances overall workflow efficiency.

The predictive maintenance facilitated by IIoT is another game-changer for manufacturing automation. In the past, equipment maintenance followed a reactive or scheduled approach, often leading to unnecessary repairs or unexpected breakdowns. IIoT systems, however, employ AI to predict when machines are likely to fail based on sensor data and historical performance. This predictive capability means that maintenance can be performed only when needed, thereby reducing costs and preventing unplanned interruptions in production. Predictive maintenance ensures that factories maintain continuous operations, delivering products to market faster and more reliably.

Quality control in manufacturing is another area profoundly impacted by automation. Traditional quality assurance processes involved manual inspection, prone to human error and limited by the inspector's subjective judgment. Automated quality control systems use machine vision and AI algorithms to inspect products down to the microscopic level, identifying defects that would be impossible for a human to see. These systems can work at high speeds, ensuring that every product meets exacting standards without compromising production rates. By implementing such advanced quality control measures, manufacturers not only improve product quality but also significantly reduce wastage.

Adaptive manufacturing, another subdomain of automation, makes production lines more flexible and responsive to changes in demand. Rather than relying on a fixed setup, adaptive systems can be reconfigured on-the-fly to produce different products or variations. This agility is achieved through smart machinery and AI-driven control systems that can adjust parameters and workflows dynamically. For instance, in the automotive industry, a single assembly line can be adapted to produce multiple car models, depending on market requirements. Such flexibility not only helps manufacturers respond to consumer demands more quickly but also significantly reduces the costs associated with changeovers.

Energy efficiency in manufacturing is also being enhanced through automation and IIoT. Traditional manufacturing processes often wasted energy due to inefficient machinery operation and lack of real-time monitoring. IIoT-enabled systems can track energy usage in real-time and optimize it by adjusting machine operation, reducing idle times, and leveraging energy-saving modes. AI algorithms can further analyze energy consumption patterns to suggest improvements. These measures contribute to significant cost savings and support sustainable manufacturing practices.

Supply chain integration is another critical benefit of manufacturing automation within the IIoT framework. By connecting manufacturing systems with ERP (Enterprise Resource Planning) and SCM (Supply Chain Management) systems, manufacturers gain comprehensive visibility into their supply chains. This integration enables just-in-time production, where components are delivered precisely when needed, minimizing inventory costs and reducing storage requirements. Additionally, real-time data exchange ensures that any disruptions in the supply chain are quickly identified and addressed, maintaining the smooth flow of materials and finished goods.

Enhanced worker safety is also a direct result of advanced manufacturing automation. Automated systems can handle hazardous tasks or operate in conditions that are unsafe for humans, such as handling toxic chemicals or working at high temperatures. Moreover, the use of AI for predictive maintenance helps prevent accidents caused by equipment failures. Health monitoring wearables connected through IIoT can further ensure that workers' safety parameters are continually assessed, providing alerts if there's any deviation from safe operating conditions. The integration of these technologies creates a safer working environment, significantly reducing the risk of workplace injuries.

The role of AI in optimizing the manufacturing process is not limited to hardware but extends to software as well. AI-driven software tools can simulate production processes, optimizing them before they are implemented. These simulations account for various factors such as material properties, machine capabilities, and environmental conditions, providing a clear picture of potential outcomes. By doing so, manufacturers can avoid costly errors and ensure that the production process is as efficient as possible from the outset.

To achieve the full potential of manufacturing automation, companies must invest in skill development and training. While automation reduces the need for manual labor in many areas, it increases the demand for highly skilled workers who can manage, operate, and maintain these advanced systems. Training programs focused on AI, robotics, and IIoT technologies are crucial for building this skilled workforce. Continuous education ensures that workers stay up-to-date with the latest advancements, enabling them to harness the full power of automated manufacturing systems.

Despite the extraordinary benefits, the road to fully automated manufacturing is not without its challenges. The initial investment in automation technology can be substantial, and the return on investment may take time to realize. Moreover, the integration of legacy systems with new IIoT technologies can present technical hurdles. However, the long-term advantages in terms of efficiency, cost savings, and market competitiveness outweigh these initial challenges. Strategic planning and phased implementation can help mitigate these issues, ensuring a smooth transition to an automated manufacturing environment.

In conclusion, manufacturing automation in the context of IIoT represents a transformative leap for the industry. Through the integration of AI and IoT, factories can operate more efficiently, safely, and flexibly than ever before. Predictive maintenance, advanced robotics, real-time data analysis, and adaptive manufacturing systems all contribute to a robust, dynamic production environment. As the technology continues to evolve, the potential for innovation within the manufacturing sector is boundless, paving the way for a future where intelligent industries drive economic growth and sustainability.

Predictive Maintenance

In the realm of Industrial IoT (IIoT), predictive maintenance stands out as a transformative application that amalgamates artificial intelligence with the interconnectedness of IoT devices. This convergence enables industries to shift from traditional maintenance practices, which are often reactive and scheduled at fixed intervals, to a more dynamic and efficient approach. By leveraging sensors, data analytics, and machine learning algorithms, predictive maintenance can foresee equipment failures before they occur, substantially reducing downtime and maintenance costs.

Imagine a vast manufacturing plant where countless machines operate continuously—each a vital cog in a larger operational framework. The traditional method often involves scheduled check-ups and maintenance, leading to either over-maintenance or unexpected breakdowns. Predictive maintenance, however, leverages data from various sensors embedded in every significant piece of equipment. These sensors track parameters like temperature, vibration, sound, and more. This real-time data is then processed by AI algorithms that can predict potential failure points and diagnose issues early on.

Let's dive deeper into how this works. First, IIoT devices capture real-time data from the machinery. This data is transmitted to a centralized system where it undergoes rigorous analysis using machine learning models specifically trained to recognize patterns indicative of equipment wear and tear or imminent failure. These models continuously learn and adapt, becoming more accurate over time. Moreover, sophisticated analytics tools provide insights and alerts, allowing maintenance teams to act proactively, thereby preventing costly downtime.

An illustrative example could be an automotive assembly line. Each robot and conveyor belt is fitted with IoT sensors that monitor its operational health. When a sensor detects an anomaly—a slight deviation in a robot's arm movement, for instance—the system flags this irregularity. The AI predicts that this deviation, if left unchecked, could lead to a significant malfunction within a week's time. Maintenance can thus be scheduled immediately to address the issue, thereby preventing a complete halt in operations.

Predictive maintenance isn't just about preventing equipment failure; it also optimizes resource allocation. Replacing parts only when necessary extends the lifespan of machinery and reduces the costs associated with unnecessary maintenance. Furthermore, it ensures safety, as equipment failure could pose serious risks to workers. By predicting and preventing such failures, companies create a safer work environment, which is invaluable in high-risk industries like mining, oil and gas, and manufacturing.

While the technology might seem complex, its benefits are straightforward: reduced unscheduled downtime, lower maintenance costs, better operational efficiency, and improved worker safety. Companies deploying predictive maintenance often see a rapid return on investment (ROI) owing to these significant operational improvements. For instance, Schneider Electric reported a 29% reduction in maintenance costs and a 20% extension in equipment longevity after implementing predictive maintenance solutions.

However, integrating predictive maintenance isn't devoid of challenges. The initial implementation requires a substantial investment in IoT devices, data infrastructure, and machine learning capabilities. Moreover, it requires a cultural shift within the organization. Traditional maintenance teams need to be trained in data analytics and machine learning concepts. There's also the complexity of managing the vast amounts of data generated by IoT devices. Companies must ensure robust data management and security practices to protect sensitive operational data from breaches.

Predictive maintenance stands as a testament to how the integration of AI and IoT can upend traditional practices, paving the way for more efficient, safer, and smarter industrial operations. As AI models become more sophisticated and IoT devices more advanced, the capabilities of predictive maintenance will only expand, making it an indispensable part of industrial operations.

In summary, predictive maintenance exemplifies the profound impact of AI and IoT convergence within the industrial sector. By enabling foresight and proactive intervention, it helps industries maintain a competitive edge, ensuring their operations remain smooth and uninterrupted. This synergy between AI and IoT doesn't just enhance the technological landscape; it revolutionizes it, setting a new standard for operational excellence.

supply chain optimization

In the context of Industrial IoT (IIoT), supply chain optimization represents a cornerstone for transforming traditional processes into highly efficient, interconnected networks. The integration of AI with IoT technologies promises a newfound level of intelligence and responsiveness that allows businesses to streamline operations, reduce costs, and increase productivity. This blend of technologies leverages real-time data from a myriad of interconnected devices, offering unprecedented visibility into every segment of the supply chain.

One of the most impressive aspects of IIoT in supply chain optimization is real-time tracking. Advanced sensors and IoT devices affixed to goods, vehicles, and storage facilities provide continuous data streams. Factories and warehouses can monitor inventory levels down to the item, receive alerts about delays, and anticipate supply chain disruptions before they escalate. This transparency helps in minimizing delays, reduces the likelihood of inventory overstocking or stockouts, and enhances the precision of inventory management—absolute game-changers in logistics.

Artificial intelligence further enhances these capabilities by applying predictive analytics to the amassed IoT data. Machine learning algorithms analyze patterns and trends, allowing supply chain managers to make informed decisions about demand forecasting, route optimization, and resource allocation. Essentially, AI shifts the paradigm from a reactive to a proactive approach. By anticipating future demand rather than merely reacting to current conditions, businesses can stay several steps ahead, ensuring smooth operation even in fluctuating market conditions.

The role of AI in mitigating risks within the supply chain cannot be overstated. With data collected from various sources, AI models can predict potential risks, such as equipment failures, adverse weather conditions, or geopolitical tensions. These predictions allow for preemptive measures, like adjusting supplier strategies or altering logistics routes, which can significantly reduce vulnerabilities. This risk management capability is especially crucial in industries where delays and disruptions can have substantial financial repercussions.

Meanwhile, IIoT also facilitates enhanced communication and collaboration across supply chain partners. Smart contracts and blockchain technologies, when integrated with IoT data, ensure transparency and trust among stakeholders. These technologies enable real-time updates and verifiable records, making it easier to manage complex supply chains that encompass multiple regions and vendors. As a result, businesses can build more resilient and integrated supply networks.

The environmental impact of supply chain optimization is another significant consideration. IIoT solutions contribute to more sustainable practices by optimizing routes to reduce fuel consumption, monitoring resources to minimize waste, and ensuring that manufacturing processes are as energy-efficient as possible. This approach not only helps in achieving compliance with environmental regulations but also enhances the public image of companies as leaders in sustainable practices.

IoT sensors also play a critical role in quality control. In industries like pharmaceuticals, food and beverage, and consumer electronics, maintaining product quality is imperative. Sensors can monitor environmental conditions such as temperature, humidity, and light exposure in real-time. Any deviations from predefined parameters trigger alerts, allowing for immediate corrective actions. This level of precision and responsiveness helps maintain product integrity from production through delivery.

Interestingly, supply chain optimization through IIoT also fosters innovation. Data collected from IoT devices provides actionable insights that can lead to process improvements, new product developments, and innovative business models. For instance, analyzing customer usage data can reveal unmet needs, prompting companies to develop new services or products that cater to those demands. Such data-driven innovation can enhance competitive advantage in the marketplace.

Furthermore, IIoT enables dynamic and adaptive supply chains. Unlike traditional systems that rely on static processes and fixed protocols, IIoT-equipped supply chains can quickly adapt to changing circumstances. For example, if a supplier faces an unexpected shutdown, IoT systems can automatically reroute orders to secondary suppliers. This adaptability not only ensures continuity but also maximizes the resilience of the supply chain.

In the context of warehouse management, IIoT drives tremendous efficiencies. Automated guided vehicles (AGVs) and robotics, powered by IoT data and AI, are revolutionizing warehouse operations. These systems can streamline the movement of goods, optimize storage space, and even handle packing and shipping tasks. The result is faster processing times, reduced labor costs, and minimized errors, all contributing to a more efficient supply chain.

Moreover, IIoT fosters greater supply chain visibility through digital twins. A digital twin is a virtual replica of physical assets, processes, or systems. By creating digital twins of supply chain components, organizations can monitor performance, predict outcomes, and optimize operations in a virtual environment before implementing changes in the real world. This technology provides a safe, cost-effective means to test and refine supply chain strategies.

Integration of IIoT in supply chains also enhances customer satisfaction. Real-time tracking allows customers to get accurate delivery estimates and status updates, improving their overall experience. This transparency builds trust and can lead to increased customer loyalty. Additionally, by optimizing supply chain operations, companies can ensure faster delivery times, fewer errors, and better service quality.

Despite these promising advancements, implementing IIoT in supply chains is not without challenges. Data security and privacy are paramount concerns. With so many interconnected devices sharing sensitive information, there is a heightened risk of cyberattacks. It is crucial for companies to invest in robust cybersecurity measures to protect their data assets and ensure the integrity of their supply chain networks.

Scalability is another pressing issue. As businesses grow and diversify, their supply chain networks become increasingly complex. IIoT solutions must be scalable to handle this complexity seamlessly. This involves not only the ability to connect and manage a growing number of devices but also ensuring that data management systems can process and analyze large volumes of information efficiently.

Interoperability between different systems and technologies is also essential for the successful implementation of IIoT. Many companies use a mix of legacy systems and modern technologies, which can lead to compatibility issues. Standardizing protocols and leveraging open-source solutions can help address these interoperability challenges, enabling smoother integration and operation of IIoT within the supply chain.

Nevertheless, the benefits of IIoT-driven supply chain optimization far outweigh the challenges. By leveraging the power of AI and IoT, businesses can create more intelligent, responsive, and resilient supply chains. These digital supply chains are not only more efficient and cost-effective but also better equipped to meet the evolving demands of the global marketplace.

As we march towards an increasingly connected future, the potential for IIoT in supply chain optimization is boundless. Companies that embrace these technologies will not only stay ahead of the curve but also set the benchmark for industry standards. The fusion of AI and IoT in supply chains signals a new era of efficiency, innovation, and sustainability—an era where the complexities of global commerce are managed with unprecedented precision and foresight.

CHAPTER 7:
AGRICULTURE AND AIoT

Integrating AI and IoT in agriculture is transforming the way we approach farming, ushering in an era of precision and efficiency unlike anything before. Using sensors and AI algorithms, precision farming optimizes crop yields by providing real-time data on soil conditions, weather patterns, and pest activities. AI-powered crop management takes it a step further by predicting potential diseases and recommending timely interventions to ensure healthy and bountiful harvests. Livestock monitoring is undergoing a revolution as well, with wearables and tracking systems enabling farmers to keep a constant watch over the health and well-being of their animals. This seamless integration of technology not only boosts productivity but also contributes to the sustainability of farming practices, offering solutions that conserve resources and minimize environmental impact. As we continue to leverage AIoT in agriculture, the potential for innovation and improvement appears boundless, promising a future where technology and nature harmoniously coexist.

Precision Farming

Precision farming, also known as precision agriculture, encapsulates the transformative power of AI and IoT in modern agriculture. This approach harnesses advanced technologies to optimize field-level management with regard to crop farming. By integrating AI algorithms and IoT devices, farmers can monitor, measure, and respond to intra-field variability in crops. This level of precision minimizes inputs, such as water and fertilizers, while maximizing output, ultimately leading to increased efficiency, sustainability, and profitability.

The essence of precision farming lies in its data-driven methodology. IoT sensors dispersed throughout agricultural fields collect vast amounts of data in real time. These sensors measure soil moisture, nutrient levels, temperature, humidity, and other environmental conditions. AI systems then analyze this data to provide actionable insights. For instance, machine learning algorithms can predict the optimal time for planting and harvesting by correlating historical weather patterns with current soil conditions.

In addition to environmental monitoring, precision farming leverages AI to enhance decision-making processes. Machine learning models can analyze past crop yields and suggest the best crop varieties for future planting. This analysis incorporates a multitude of factors, including soil type, climatic conditions, and pest prevalence, effectively aiding farmers in selecting the most suitable crops for their land. Moreover, AI-driven analytics can identify disease outbreaks at an early stage, enabling preemptive measures to mitigate potential yield losses.

Precision farming also redefines irrigation practices through smart water management systems. Traditional irrigation methods often lead to water wastage, either through over-irrigation or inefficient distribution. IoT sensors combined with AI analytics can assess soil moisture levels across different zones of a field and activate irrigation systems only where and when needed. This targeted approach not only conserves water but also ensures that crops receive optimal hydration.

Another critical area where precision farming excels is in fertilizer application. Excessive use of fertilizers not only increases costs but also poses environmental risks, such as soil degradation and water pollution. Precision farming techniques enable variable rate technology (VRT), where fertilizers are applied variably across a field based on precise nutrient requirements determined through soil sensors and AI analysis. This ensures that each section of the field receives exactly what it needs, reducing wastage and environmental impact.

The scope of precision farming extends to pest and disease management. IoT-based sensors can continuously monitor crop health, while AI algorithms analyze patterns to detect early signs of pest infestations or disease outbreaks. Predictive analytics can then forecast potential threats, allowing farmers to take timely preventive actions. For example, drones equipped with multispectral cameras can cover large areas quickly, capturing detailed images that AI systems analyze for signs of stress or pest presence. This timely intervention helps in minimizing crop damage and maintaining healthy yields.

Precision farming is not just about in-field operations. It encompasses a broad spectrum of farm management activities, including machinery optimization. GPS-guided tractors and drones ensure accurate planting, weeding, and harvesting, reducing the overlap and gaps that occur with manual operations. AI-powered machinery can also self-diagnose issues, predict maintenance needs, and even adapt to varying field conditions on the go. Such advancements lead to more efficient use of resources, lower operational costs, and improved crop production.

While the integration of AI and IoT in precision farming presents numerous advantages, it also entails certain challenges. Data management is a significant hurdle; the sheer volume of data generated by IoT sensors can be overwhelming. Ensuring data accuracy and integrity is crucial for reliable analytics. Additionally, the cost of implementing precision farming technologies can be a barrier for small-scale farmers. However, innovation in these areas continues to lower costs and make the technology more accessible.

Furthermore, precision farming necessitates a paradigm shift in how farmers approach agriculture. The traditional experience-based methods need to be complemented by technology-driven insights.

This calls for training and education, ensuring that farmers can effectively leverage these advanced tools. Many organizations are stepping up to provide resources and support, bridging the gap between technology and traditional farming practices.

The environmental benefits of precision farming are undeniable. By optimizing inputs, such as water and fertilizers, it promotes sustainable agricultural practices. Reducing chemical run-off and conserving water resources align with global efforts to combat climate change and preserve valuable natural ecosystems. Precision farming also contributes to food security by enhancing crop yields and ensuring consistent production, even as environmental conditions become more unpredictable due to climate change.

In the context of the broader agricultural landscape, precision farming represents a significant step towards digital transformation. As AI and IoT technologies continue to evolve, their applications will become more sophisticated, further enhancing the capabilities of precision farming. The fusion of real-time data analytics with smart machinery and advanced algorithms will drive the next wave of agricultural productivity and sustainability.

In summary, precision farming symbolizes the intersection of technology and agriculture, where AI and IoT collaboratively pave the way for a more efficient, sustainable, and productive future. This methodology not only transforms traditional farming practices but also ensures that agriculture can meet the demands of a growing global population. By embracing precision farming, we are setting the stage for a new era of agricultural excellence, characterized by innovation, efficiency, and sustainability.

AI-Powered Crop Management

As agriculture continues to evolve, the integration of AI and IoT, or "AIoT," is unlocking significant potential in crop management.

AI-powered crop management combines advanced data analytics, machine learning algorithms, and IoT devices to optimize every stage of the agricultural process—from planting to harvesting. This fusion of technology is not just a futuristic concept. It's actively transforming how farmers manage crops, making agriculture more efficient, sustainable, and productive.

One of the key aspects of AI-powered crop management is precision agriculture. Precision agriculture leverages data from various sensors placed throughout the farm to monitor conditions such as soil moisture, nutrient levels, and crop health in real-time. Using this data, AI algorithms can make precise recommendations for irrigation, fertilization, and pest control, eliminating guesswork and reducing waste.

For instance, imagine a farmer who traditionally irrigates an entire field uniformly. With AI-powered crop management, sensors collect soil moisture data at multiple points across the field. The AI system analyzes this data and determines the specific irrigation needs of different areas. By doing so, water usage is optimized, leading to better crop yields and significant water conservation.

Nitrogen, phosphorus, and potassium levels—essential nutrients for plants—can be managed more effectively with AIoT solutions. Drones equipped with multispectral cameras can capture detailed images of crop fields. AI models then interpret these images to assess nutrient deficiencies and suggest targeted fertilization strategies. Rather than applying a blanket amount of fertilizer across the entire field, farmers can apply just the right amount where it's needed the most, minimizing cost and environmental impact.

Pest and disease management is another crucial area where AI-powered systems show immense promise. Traditional methods often involve widespread pesticide application, which can be harmful to the environment and non-target organisms. AIoT systems can proactively identify potential pest outbreaks. In situ sensors and image recognition AI can detect early signs of pest infestation or disease, allowing for timely and localized intervention. This approach not only enhances crop protection but also reduces the reliance on chemical pesticides, promoting more sustainable farming practices.

Moreover, AI-powered crop management systems can predict and respond to weather patterns. By analyzing historical weather data alongside real-time meteorological information, AI models can forecast adverse weather conditions, such as droughts or heavy rainfall. These predictions enable farmers to take preemptive measures, like adjusting irrigation schedules or reinforcing plant cover, to mitigate potential damage to crops.

One practical example of this technology in action is in the vineyard industry. Grapevines are highly sensitive to changes in climate and soil conditions. AI-powered management systems can monitor indicators such as temperature, humidity, and soil pH, providing vintners with real-time insights. These insights help in making informed decisions on irrigation, pruning, and harvesting times, ultimately improving grape quality and yield.

Additionally, AI-powered crop management aids in optimizing harvest schedules. Through continuous monitoring and predictive analytics, AI systems can determine the optimal time for harvesting crops. This automation ensures that crops are harvested at their peak, maximizing both yield and quality. In regions growing perishable goods like fruits and vegetables, timely harvests are crucial for minimizing post-harvest losses and ensuring the freshness of the produce.

On a broader scale, AI-powered crop management is fostering a shift towards data-driven agriculture. IoT devices like weather stations, soil sensors, and GPS-enabled machinery generate vast amounts of data that feed into AI systems. These systems process the data to provide actionable insights through user-friendly interfaces accessible on smartphones or computers. Farmers can access comprehensive reports and recommendations, enabling them to make well-informed decisions quickly and efficiently.

The automation aspect of AI-powered crop management also extends to machinery. Autonomous tractors and robotic harvesters, guided by AI algorithms, are revolutionizing field operations. These machines can plant seeds with precision, apply fertilizers, and harvest crops with minimal human intervention. The integration of AI ensures that these tasks are performed with high accuracy, improving efficiency and reducing labor costs.

Another significant benefit of AIoT in agriculture is its potential to enhance supply chain management. By synchronizing with IoT-enabled logistics systems, farmers can track the journey of their produce from field to market. This real-time monitoring helps in ensuring that the produce remains within optimal conditions, reducing spoilage and waste. Enhanced traceability also adds an extra layer of accountability and transparency, boosting consumer trust and marketability.

From a sustainability perspective, AI-powered crop management contributes to more eco-friendly farming practices. By optimizing resource use—such as water, fertilizers, and pesticides—these systems reduce the ecological footprint of agriculture. Efficient resource utilization not only conserves the environment but also aligns with the growing consumer demand for sustainably produced food.

Collaboration between tech companies and agricultural experts is crucial for advancing AI-powered crop management. Many tech startups and established agri-tech firms are working together to develop and deploy these solutions. These collaborations are fostering innovation, bringing cutting-edge technology to the forefront of agriculture, and making AI-powered crop management accessible to farmers worldwide.

Education and training for farmers are essential components of adopting AIoT solutions. Many farmers may lack the technical know-how to operate advanced AI-powered systems. Initiatives to educate and train farmers in the use of these technologies are essential for broader adoption. Workshops, online courses, and support services can empower farmers to leverage AIoT effectively, enhancing their productivity and sustainability.

Challenges do exist, however. Data privacy and security remain significant concerns, especially given the vast amount of data generated and shared. Ensuring that this data is protected from cyber threats and unauthorized access is paramount. Moreover, the initial cost of implementing AIoT systems can be prohibitive for small-scale farmers. There is a need for policy interventions and financial support to make these technologies affordable and inclusive.

Despite these challenges, the future of AI-powered crop management looks promising. Emerging technologies like blockchain can further enhance transparency and traceability in the agricultural supply chain. Machine learning models will continue to evolve, becoming more accurate and capable of handling diverse agricultural scenarios. Integration with other cutting-edge technologies like drones and satellite imagery will provide even more granular insights, making crop management smarter and more responsive.

In conclusion, AI-powered crop management is set to revolutionize agriculture. By harnessing the power of AI and IoT, farmers can achieve unprecedented levels of efficiency, productivity, and sustainability. This technological advancement is not just about enhancing crop yields; it represents a paradigm shift towards smarter, data-driven farming practices that benefit both the farmer and the environment. As we continue to innovate and refine these systems, the potential for positive impact on global agriculture is immense, paving the way for a future where farming is more intelligent, responsive, and sustainable than ever before.

Livestock Monitoring

The integration of artificial intelligence (AI) and the Internet of Things (IoT), often referred to as AIoT, is bringing transformative changes to livestock monitoring within the agricultural sector. Traditionally, livestock management relied heavily on manual labor, with farmers tracking the health, location, and overall well-being of their animals through physical inspections and basic record-keeping. Today, AIoT provides a host of sophisticated tools, enhancing efficiency, accuracy, and productivity while paving the way for more sustainable and ethical farming practices.

AI and IoT technologies work harmoniously to create smart environments in which real-time data collection and analysis become the norm. This synergy allows farmers to monitor livestock in real-time using wearable sensor devices attached to the animals. These devices collect a wide range of data, from physical activity and location to health indicators like temperature and heart rate. This continuous stream of data provides farmers with a level of insight that was previously unattainable.

One significant benefit of this technological shift is the improvement in animal health management. With AI-powered algorithms analyzing the data collected from IoT devices, unusual patterns can be detected early, alerting farmers to potential health issues before they become serious. This predictive capability helps in the timely administration of treatments, preventing disease outbreaks and reducing the mortality rate among livestock. It also means a reduction in the use of antibiotics, supporting the fight against antibiotic resistance.

Additionally, AIoT solutions aid in optimizing feeding practices. Smart feeding systems use data from sensors to determine the right amount and type of feed each animal requires. This customized feeding approach ensures that the nutritional needs of individual animals are met, promoting better growth and productivity while minimizing waste. These efficiencies are not only beneficial from an economic standpoint but also contribute to environmental sustainability by reducing the resource inputs required for livestock farming.

The advent of AIoT in livestock monitoring also enhances breeding programs. Genetic improvements and selective breeding are critical to increasing yield and resilience in livestock populations. Through AI analysis, farmers can identify the best candidates for breeding based on a comprehensive set of health and performance indicators. Such an analytical approach ensures that breeding decisions are data-driven and aimed at achieving long-term gains in productivity and health.

Moreover, GPS-enabled tracking systems facilitate better herd management by monitoring the movement and location of livestock. This is particularly valuable for large farms where keeping an eye on every animal is an overwhelming task. If an animal strays too far from the designated area or displays unusual movement patterns, alerts are sent to the farmer. This not only helps in preventing loss and theft but also ensures the animals' well-being by allowing quick interventions in case of accidents or distress.

Incorporating AIoT within livestock monitoring isn't without its challenges. Connectivity issues in remote agricultural areas can pose significant hurdles, as consistent data transmission is crucial for real-time monitoring. To address this, advancements in telecommunications infrastructure are vital. Solutions such as low-power wide-area networks (LPWAN) and satellite internet can provide reliable connectivity even in the most isolated regions, ensuring that the benefits of AIoT are widely accessible.

Data privacy and security also emerge as critical issues with the proliferation of AIoT devices. The vast amount of data generated needs to be stored securely and used responsibly. Implementing strong cybersecurity measures and ensuring compliance with data protection regulations are essential steps in safeguarding the information collected. Farmers must be educated about these aspects to foster trust and promote the widespread adoption of AIoT technologies.

AIoT-driven livestock monitoring offers promising economic incentives as well. By reducing the costs associated with disease management, feed optimization, and manual labor, these technologies make farming more profitable. Increased productivity and efficiency translate into higher yields and better market returns. As the agricultural industry becomes increasingly competitive, those integrating advanced technologies into their operations are likely to gain a significant edge.

However, the shift toward AIoT in livestock monitoring isn't solely about economic gains. It also presents a moral and ethical dimension. Better monitoring systems ensure that animals are treated more humanely, with their needs being promptly and accurately addressed. This aligns with the growing consumer demand for ethically produced food products. By adopting AIoT technologies, farmers demonstrate a commitment to animal welfare, enhancing their reputation and gaining consumer trust.

Community and governmental support play pivotal roles in encouraging the adoption of AIoT in livestock monitoring. Programs offering financial incentives and technical assistance can accelerate the transition. Collaborative efforts among stakeholders, including technology providers, agricultural extension services, and research institutions, are essential in developing practical and scalable solutions tailored to farmers' needs.

To maximize the benefits of AIoT in livestock monitoring, ongoing research and innovation are imperative. As technology evolves, new applications and improved versions of existing tools will emerge. Continuous learning and adaptation will ensure that farmers remain at the forefront of technological advancement. Investing in training programs that equip farmers with the knowledge and skills to effectively use AIoT technologies is crucial for sustained progress.

Looking to the future, the potential of AIoT in livestock monitoring is boundless. As artificial intelligence becomes more sophisticated and sensing technology advances, the precision and scope of monitoring capabilities will expand even further. Enhanced integration with other agricultural systems will facilitate a holistic approach to farm management, driving the industry toward a new era of smart farming.

In summary, the integration of AIoT in livestock monitoring is revolutionizing the agricultural industry. It provides comprehensive and real-time insights into livestock health, feeding, breeding, and movement, resulting in increased productivity, reduced costs, and enhanced animal welfare. While challenges such as connectivity and data security need addressing, the benefits far outweigh the hurdles. Embracing these technologies promises a future where farming is more efficient, sustainable, and humane, ultimately shaping the future of agriculture in profound and positive ways.

CHAPTER 8:
AI AND IoT IN RETAIL

In the rapidly evolving retail landscape, the integration of AI and IoT technologies is revolutionizing customer experiences, streamlining inventory management, and optimizing supply chain logistics. Imagine walking into a store where AI-driven insights predict your preferences, offering personalized recommendations while IoT-enabled sensors track inventory levels in real-time, ensuring popular items are always in stock. This seamless blend of AI and IoT not only enhances operational efficiency but also fosters a more engaging and responsive shopping environment. Retailers leveraging these advanced technologies can anticipate consumer demands with unparalleled accuracy, reduce waste through predictive stocking, and create a supply chain that is both agile and resilient. The transformative power of AI and IoT in retail is not just about staying ahead of the competition but about redefining how businesses connect with their customers in more meaningful and intelligent ways.

Customer Experience Enhancement

In today's retail landscape, the blending of Artificial Intelligence (AI) and the Internet of Things (IoT) is creating a profound transformation in the customer experience. Traditionally, shopping was a relatively straightforward ordeal. You'd walk into a store, browse the shelves, and purchase what you needed. However, the advent of AI and IoT has turned this simplistic model on its head, making shopping an interactive, personalized, and highly efficient process.

At the heart of this transformation is data. More specifically, it's the ability of AI to process vast amounts of data quickly and accurately. For instance, consider predictive analytics. By analyzing past purchasing behavior, social media activity, and even weather patterns, AI can predict what products a customer is likely to purchase next.

Meanwhile, IoT collects real-time data from various endpoints, such as smart shelves and connected devices, providing an almost instantaneous feedback loop. When these technologies converge, retailers can create highly personalized shopping experiences that feel almost bespoke to individual customers. The objective is to meet and even exceed customer expectations by delivering products and recommendations tailored specifically to their preferences and needs.

One of the most impressive manifestations of this technology duo in retail is the concept of the "smart store." In these stores, sensors and cameras powered by AI analyze foot traffic patterns, shelf interactions, and even facial expressions. This data allows stores to optimize layouts, restock popular items faster, and minimize bottlenecks, all contributing to a smoother, more enjoyable shopping experience. For example, smart shelves can detect when an item is running low and trigger an automatic restock alert, ensuring popular products are always available.

Moreover, virtual assistants and chatbots have become sophisticated customer service tools. These AI-driven entities can answer queries, handle complaints, and even complete transactions around the clock, providing customers with instant support. The seamless integration of these tools into mobile apps and e-commerce platforms means that help is always just a click away. This level of support increases customer satisfaction tremendously, as issues are resolved more quickly and efficiently.

Shopper loyalty programs have also undergone a significant transformation thanks to AI and IoT. By tying customer rewards to a holistic profile powered by AI analytics, retailers can offer highly personalized rewards. For instance, if data indicates that a customer frequently purchases organic products, they might receive a special discount on a new line of organic goods. These individualized incentives help foster a deeper connection between the brand and the customer.

The role of augmented reality (AR) cannot be overlooked in enhancing the customer experience. With AI and IoT, AR can be used to enrich the shopping journey substantially. Consider an app that allows customers to visualize how a piece of furniture would look in their living room or see how a certain shade of lipstick complements their skin tone before making a purchase. These interactive tools not only make shopping more fun but also reduce the likelihood of returns, as customers are more confident in their purchases.

IoT plays a significant role behind the scenes as well. Smart tags and IoT-enabled devices track products through every stage of the supply chain. This not only ensures that shelves are stocked efficiently but also provides customers with precise delivery times and real-time updates on order status. Imagine receiving a notification on your phone that the exact pair of shoes you've been eyeing for weeks is back in stock at a local store. This real-time interaction is where IoT shines in enhancing the customer experience.

Personalized marketing through AI-driven insights is another avenue retailers are exploring. AI algorithms analyze browsing history, purchase patterns, and even online behavior to create hyper-targeted advertising campaigns. These campaigns are delivered through multiple channels, including email, social media, and push notifications, ensuring that the message reaches the customer where they are most likely to engage. This form of targeted marketing can be incredibly effective, as it caters directly to individual interests and needs.

Digital wallets and contactless payment systems, supported by IoT, are also reshaping the retail environment. Customers can now make payments faster and more securely than ever before, often without needing to touch anything beyond their own mobile device. These systems integrate seamlessly with AI to offer tailored financial services and spending insights, further enhancing the customer's financial well-being and shopping experience.

But it's not just about the transactional phase of shopping. AI and IoT are revolutionizing the way customers discover products. AI-driven recommendation engines can present customers with items they didn't even know they wanted, based on complex algorithms that analyze their behavior and preferences. Similarly, IoT devices like smart mirrors in clothing stores offer customers recommendations on what to try next, based on real-time data about inventory and style trends.

These technologies also offer valuable feedback for continuous improvement. By collecting and analyzing customer data, retailers can identify pain points and areas for enhancement. For instance, if data shows a high rate of cart abandonment at a specific point in the checkout process, AI can analyze the factors contributing to this issue and suggest solutions to streamline the process. This could mean simplifying form fields, offering additional payment options, or even initiating a chatbot to assist in real-time.

Furthermore, loyalty programs are experiencing a renaissance. Gone are the days of generic, blanket rewards. Today, AI-driven loyalty programs analyze an array of customer data to provide personalized offerings. By leveraging machine learning, these programs adapt over time, learning more about each customer and refining the rewards to become ever more appealing. This not only incentivizes repeat business but also strengthens the brand-customer relationship.

The impact of AI and IoT on retail isn't limited to the interaction between customers and products; it extends to the entire retail environment. For example, smart energy management systems use IoT to monitor and optimize the store's energy consumption in real-time, creating more comfortable shopping environments while reducing costs. These savings can then be passed on to customers, creating a more favorable price point.

Visual merchandising has also evolved with these technologies. AI can analyze customer flow and engagement with displays to determine the most effective arrangements. IoT sensors can detect movement and infer shopper interest in certain products, providing invaluable data for maximizing the attractiveness and effectiveness of retail displays. By continuously iterating on these insights, retailers create visually stunning and effective displays that drive both engagement and sales.

Additionally, AI and IoT contribute to social responsibility and ethical business practices, which are increasingly important to today's consumers. By optimizing supply chains and reducing waste, these technologies help retailers operate more sustainably. Customers are more likely to support businesses that align with their values, and AI and IoT technologies facilitate a level of transparency and efficiency that wouldn't be possible otherwise.

Even the in-store experience can be dramatically improved. Smart fitting rooms equipped with IoT devices allow customers to quickly request different sizes or items without leaving the fitting room. AI-driven virtual assistants can suggest complementary products based on what a customer is trying on, thereby increasing upsell opportunities while enhancing the shopping experience.

Of course, all these improvements hinge on the ethical use and management of customer data. Trust is fundamental to the relationship between retailers and their customers. Responsible implementation of AI and IoT ensures that data is handled with utmost care, transparency, and

Inventory Management

The integration of AI and IoT technology in retail is fundamentally transforming the way businesses manage their inventories. Traditionally, inventory management relied heavily on periodic manual counts, static spreadsheets, and human oversight, which often led to inefficiencies such as overstocking, understocking, and significant financial losses due to mismanagement. Today's landscape, however, is dramatically different. AI and IoT enable real-time visibility and predictive analytics, guiding retailers towards a more data-driven approach.

One major advantage of AI-driven inventory management is predictive analytics. Through machine learning algorithms, AI can analyze historical sales data and identify patterns that humans might miss. It can predict demand for different products in varying seasons, regions, and even down to specific store locations. For example, a retailer might stock up on umbrellas based on weather forecasts and past precipitation data collected via IoT sensors. The predictive capabilities of AI help in maintaining optimal stock levels, reducing the risk of lost sales due to stockouts or excess inventory sitting idle on shelves.

In parallel, IoT solutions bring an additional layer of real-time tracking and automation. RFID tags, smart shelves, and connected sensors provide instant updates on inventory levels, item locations, and even conditions such as temperature and humidity for perishable goods. This continuous monitoring allows for immediate response to discrepancies, further ensuring that inventory levels remain accurate and up-to-date. Inventory counts that once took hours or even days to perform manually can now be completed in real-time, freeing up staff to focus on more strategic tasks.

Moreover, AI-enhanced demand forecasting is crucial in a dynamic market. Traditional forecasting methods often struggled to adapt quickly to sudden changes in consumer behavior or external conditions. By contrast, AI can process vast amounts of diverse data—social media trends, economic indicators, local events, and even cross-channel sales data—enabling it to adapt forecasts in near re-al-time. This agility proves invaluable in responding to unexpected events, such as during the holiday season or in the wake of sudden market trends.

Another transformative aspect is automated replenishment. With AI and IoT working in tandem, the system can automatically trigger orders when inventory levels fall below a certain threshold. This reduces the dependency on human intervention and the associated risks of delays or oversight. For example, a smart shelf equipped with IoT sensors can detect low stock levels and immediately notify an AI system that processes the data to place an order. This seamless automation mitigates the chances of stockouts and ensures a consistent supply chain flow.

AI's role doesn't end at predictive analytics and automated replenishment. It also extends to optimizing warehouse operations. Machine learning algorithms can analyze the movement of goods within a warehouse and suggest the most efficient layouts. They can identify slow-moving items and reposition them to less accessible areas while placing high-demand products in easily accessible spots. Automated guided vehicles (AGVs) and robotic picking systems further streamline the inventory management process by reducing the time and labor costs associated with manual picking and packing.

IoT devices also contribute significantly to improving accuracy and reducing shrinkage. Smart cameras and sensors can monitor inventory in real-time and flag any suspicious activities. For instance, in a high-value electronics store, IoT-enabled surveillance systems can track each item's journey from the backroom to the shelf and eventually to the point of sale. Any discrepancies can be quickly addressed, thus minimizing theft and other forms of shrinkage. This level of visibility ensures not only the security of the products but also fosters trust among stakeholders.

The synergy between AI and IoT also extends to improving the supply chain's efficiency. Enhanced logistics platforms powered by AI can optimize routes for delivery trucks, taking into account factors such as traffic conditions, fuel efficiency, and delivery windows. IoT devices within these vehicles provide real-time updates on their locations, speeding up the delivery process and ensuring timely restocking. Retailers can thus maintain a lean inventory, reducing overhead costs and wastage, while still meeting customer demands swiftly.

Real-time inventory tracking enabled by AI and IoT also facilitates better decision-making at the managerial level. Instead of relying on periodic reports, managers now have access to real-time dashboards displaying inventory levels, sales trends, and other critical metrics. This real-time visibility empowers them to make informed decisions quickly, whether it's to stock up on a best-selling item or discontinue a slow-moving product. The ability to act on up-to-the-minute data provides a significant competitive advantage.

Moreover, customer satisfaction sees a boost with more efficient inventory management. When stock levels are managed accurately, customers are less likely to experience the frustration of finding out that a desired item is out of stock. Enhanced inventory management ensures better product availability, contributing to a smoother shopping experience. This is especially critical in an age where customer loyalty can be fickle and the next competitor is just a click away.

Beyond physical stores, e-commerce platforms also benefit immensely from AI and IoT-driven inventory management. Online retailers often face the challenge of integrating multiple sales channels while maintaining a coherent inventory system. AI can unify inventory data from brick-and-mortar stores, online platforms, and even third-party logistics providers. This holistic view prevents overselling and ensures that customers receive accurate information about product availability, thus enhancing their online shopping experience.

AI and IoT also facilitate cross-channel inventory management by providing a unified view of stock levels across different retail formats. This ensures that if an item is out of stock in one store, it can be sourced from another location or even directly from the warehouse, enhancing the overall customer experience. This level of integration is crucial in an omnichannel retail environment, where customers expect a seamless experience whether they shop online, in-store, or through a mobile app.

Looking ahead, the potential for AI and IoT in inventory management is virtually limitless. As technology continues to evolve, the integration of more advanced AI capabilities and more sophisticated IoT devices will only enhance these benefits. Innovations like edge computing can further streamline processes by enabling data processing closer to the source, reducing latency, and improving real-time decision-making. Additionally, as quantum computing becomes more accessible, the computational power available for AI algorithms will skyrocket, leading to even more accurate predictive models.

However, it's essential to acknowledge the challenges that come with the implementation of AI and IoT technologies in inventory management. Data security remains a significant concern, as the increased reliance on interconnected devices and cloud storage presents more points of vulnerability. Retailers must adopt robust cybersecurity measures to protect sensitive information. Another challenge is the initial investment required for AI and IoT infrastructure. While the long-term benefits often justify the cost, smaller retailers might struggle with the upfront expenses. Nevertheless, as the technology becomes more widespread and accessible, these barriers are likely to diminish over time.

In conclusion, the integration of AI and IoT in inventory management is revolutionizing the retail industry. By leveraging predictive analytics, real-time tracking, and automated replenishment, retailers can optimize their stock levels, enhance operational efficiency, and provide a superior customer experience. As these technologies continue to evolve, the boundaries of what's possible will expand even further, paving the way for a future where inventory management is not just an operational necessity but a strategic advantage.

Supply chain and logistics

AI and IoT are fundamentally reshaping the landscape of supply chain and logistics in retail. The traditional paradigms of supply chain management are evolving, driven by the promise of enhanced visibility, efficiency, and predictive capabilities. In essence, we're witnessing a transformation that spans from the initial stages of raw material procurement to the final delivery of products to consumers.

One of the most profound impacts is the enhancement in supply chain visibility. By integrating IoT sensors throughout the supply chain, companies can monitor the status and location of goods in real-time. This transparency is invaluable, allowing businesses to track shipments, monitor inventory levels, and identify potential bottlenecks before they become problematic. Imagine a refrigerated truck transporting perishable goods; IoT sensors can monitor temperature and humidity conditions continuously, ensuring that the products remain within the specified parameters. If any deviation is detected, alerts can be immediately sent to the concerned stakeholders, mitigating potential spoilage.

Artificial Intelligence takes this visibility a step further by enabling predictive analytics. Machine learning algorithms analyze the vast amounts of data generated by IoT sensors to forecast demand, optimize routes, and predict maintenance needs. For instance, AI can predict when a specific machine along the supply chain is likely to fail, allowing for preemptive maintenance, reducing downtime, and avoiding costly disruptions. Similarly, demand forecasting powered by AI helps retailers ensure they have the right products at the right time, minimizing overstock and stockouts.

The combination of AI and IoT is also revolutionizing inventory management. Traditional inventory methods often involve periodic manual checks that are time-consuming and prone to errors. In contrast, IoT-enabled warehouses use sensors and RFID tags to automate inventory tracking. These sensors update inventory counts in real-time, reducing human error and providing accurate stock levels instantaneously. AI then steps in to analyze this data, identifying trends and patterns that inform inventory replenishment strategies. This seamless integration of AI and IoT ensures that retailers maintain optimal inventory levels, reducing holding costs and improving cash flow.

Logistics optimization is another area where AI and IoT are making significant strides. By analyzing route data, traffic conditions, and delivery schedules, AI algorithms can generate the most efficient routes for delivery trucks, reducing fuel consumption and ensuring timely deliveries. IoT devices can provide real-time updates on traffic conditions, allowing dynamic rerouting to avoid delays. This not only enhances operational efficiency but also improves customer satisfaction by providing accurate delivery timelines.

Moreover, the integration of AI and IoT is enabling more sustainable supply chain practices. Optimized routing and efficient inventory management directly contribute to reduced carbon footprints. For instance, by minimizing the distance traveled and avoiding unnecessary trips, companies can significantly cut down on fuel usage and emissions. Smart packaging solutions enabled by IoT also play a role in sustainability, as intelligent sensors can monitor the condition of products, reducing waste caused by spoilage or damage during transit.

Supplier relationship management has also seen improvements with the advent of AI and IoT. Companies can use data collected from IoT sensors to monitor supplier performance closely. AI algorithms analyze this data to provide insights into supplier reliability, delivery times, and quality metrics. Armed with this information, retailers can negotiate better terms, identify and build relationships with high-performing suppliers, and address issues proactively before they escalate into significant problems.

One can't overlook the role of blockchain technology in conjunction with AI and IoT in supply chain and logistics. Blockchain provides a decentralized, immutable ledger that ensures data transparency and security. When IoT devices capture data at each touchpoint of the supply chain, it can be recorded on a blockchain, creating a verifiable trail from manufacturer to consumer. AI can then analyze this blockchain data to identify inefficiencies, trace the origin of defects, and enhance overall supply chain integrity.

AI-driven autonomous vehicles and drones are another manifestation of this technological synergy in logistics. Autonomous delivery vans and drones equipped with IoT devices can handle last-mile delivery with minimal human intervention. These smart vehicles use AI to navigate traffic and optimize routes while their IoT devices provide real-time location updates and condition monitoring. This not only speeds up the delivery process but also reduces labor costs and the potential for human error.

While the adoption of AI and IoT in supply chain and logistics promises significant benefits, it also comes with its share of challenges. Data security is paramount given the vast amounts of sensitive data being generated and transmitted across the network. Ensuring robust cybersecurity measures to protect this data from breaches and unauthorized access is crucial. Additionally, the integration of legacy systems with new IoT devices and AI algorithms can be complex and may require substantial investment in terms of time and resources.

Furthermore, there are ethical considerations surrounding the deployment of these technologies. For instance, the implementation of AI and IoT could potentially displace certain jobs, particularly those involving manual, repetitive tasks. It's essential for businesses to address these concerns by investing in reskilling and upskilling their workforce, ensuring that employees are equipped to take on more strategic, value-added roles.

Despite these challenges, the benefits of integrating AI and IoT into supply chain and logistics far outweigh the potential downsides. Companies that embrace these technologies are better positioned to navigate the complexities of the modern retail landscape, delivering superior value to their customers while maintaining operational excellence.

Ultimately, the fusion of AI and IoT in supply chain and logistics is more than just a technological advancement; it's a strategic imperative for retailers aiming to stay competitive in an increasingly dynamic market. By leveraging the power of these technologies, businesses can create a more responsive, efficient, and sustainable supply chain that not only meets but exceeds the expectations of today's discerning consumers.

CHAPTER 9:
ENERGY AND UTILITIES

The integration of AI and IoT in energy and utilities is set to revolutionize how we manage and utilize resources, making systems smarter, more efficient, and highly responsive. Imagine power grids that not only distribute electricity but also anticipate and adjust to real-time demand fluctuations, driven by AI algorithms that optimize energy distribution down to the last watt. Renewable energy sources, previously volatile due to their dependency on weather conditions, become more reliable, as intelligent systems predict and adapt to changes, ensuring maximum output and minimal waste. Furthermore, efficient resource management extends beyond electricity, encompassing water and gas utilities that are monitored and regulated through interconnected devices. These advancements don't just promise operational efficiencies; they also pave the way for more sustainable practices, reducing carbon footprints and conserving essential resources for future generations. The synergy of AI and IoT in this sector is not just a glimpse into the future but a palpable shift already in motion, promising to redefine the landscape of energy and utilities.

Smart Grids

The integration of AI and IoT into the energy sector has revolutionized traditional utilities, giving rise to what we now term "Smart Grids." A Smart Grid is essentially a modernized electrical grid that uses digital communication technology to detect and react to local changes in usage. This technology improves the reliability, efficiency, and sustainability of electricity services.

One of the primary benefits of Smart Grids is their capacity to manage electricity demand more effectively. By employing AI algorithms and IoT sensors, these grids can predict electricity consumption patterns and adjust the distribution accordingly. This not only minimizes energy waste but also stabilizes the grid, especially during peak demand periods. Imagine a hot summer day when everyone turns on their air conditioners—Smart Grids can dynamically optimize power flow, ensuring no single area is overwhelmed.

Another crucial aspect is the ability to integrate renewable energy sources seamlessly. Whether it's solar panels, wind turbines, or hydropower plants, Smart Grids use AI to predict the amount of energy generated and balance it with the grid's current demand. This reduces reliance on fossil fuels, cuts greenhouse gas emissions, and moves us closer to a sustainable energy future. The integration with renewable energy doesn't just stop at large-scale units; even residential solar panels can contribute to the grid's overall supply, thanks to Smart Grid technology.

Moreover, Smart Grids enhance the reliability of the energy supply. Traditional power grids are susceptible to blackouts and failures, but Smart Grids utilize IoT devices to monitor the grid's health in real-time. AI systems can detect anomalies and predict potential failures before they occur, triggering preventative maintenance. For utility companies, this diagnostic approach significantly reduces downtime and maintenance costs.

Smart Grids also offer consumer benefits, allowing for greater transparency and control over energy usage. Advanced metering infrastructure (AMI) systems provide real-time feedback on electricity consumption, enabling consumers to make informed choices to reduce their energy bills. Additionally, utility companies can implement variable pricing models, incentivizing consumers to use electricity during off-peak hours, thereby flattening demand peaks and conserving resources.

The fusion of AI and IoT within Smart Grids empowers utilities to operate more autonomously and efficiently. Through machine learning algorithms, these grids continuously learn and optimize their performance, adapting to new patterns of electricity generation and consumption. While the infrastructure and initial investment may be significant, the long-term benefits, both in terms of operational efficiency and environmental impact, are substantial.

On a larger scale, Smart Grids contribute to the resilience of cities and communities. By decentralizing energy generation and distribution, cities become less vulnerable to large-scale grid failures. Microgrids—a smaller version of the Smart Grid—can operate independently from the central grid, providing localized energy solutions for neighborhoods or specific facilities. These microgrids are particularly valuable in disaster recovery scenarios, where conventional grids may be compromised.

Regulatory frameworks and policies play a significant role in the deployment and success of Smart Grids. Governments and regulatory bodies must support the transition by enacting laws and guidelines that promote innovation while ensuring the secure and equitable distribution of resources. Investment in research and development, as well as public-private partnerships, will be key to overcoming the initial barriers to widespread Smart Grid adoption.

From an economic standpoint, Smart Grids can stimulate job creation and drive technological advancements. The shift towards smarter, cleaner energy systems requires a skilled workforce adept in AI, IoT, and cybersecurity. As Smart Grid technology evolves, so do the opportunities for innovation within the energy sector, leading to the growth of new industries and economic models.

Lastly, Smart Grids have a profound societal impact. They democratize energy access by making it easier to harness and distribute renewable energy sources, which can be particularly beneficial for underserved and remote communities. By bridging the energy divide, Smart Grids contribute to social equality and improve the quality of life for all individuals.

The path to fully realizing the potential of Smart Grids is fraught with challenges, from technical obstacles to regulatory uncertainties. However, the ongoing advancements in AI and IoT, coupled with a collaborative effort from stakeholders, promise a future where Smart Grids could very well be the norm rather than the exception. Embracing such innovations paves the way for a more sustainable, reliable, and efficient energy ecosystem.

In conclusion, Smart Grids stand at the forefront of the energy revolution brought about by AI and IoT. Through intelligent management of energy resources, they offer significant advantages—ranging from enhanced reliability and efficiency to improved sustainability and consumer control. As the technology continues to evolve, Smart Grids will undoubtedly be a cornerstone in the foundation of a smarter, greener future.

AI in Renewable Energy

Renewable energy sources, such as solar, wind, hydro, and geothermal, present unique challenges in terms of efficiency and reliability. Integrating Artificial Intelligence (AI) can address these challenges by optimizing energy production, predicting maintenance needs, and managing the grid more effectively. AI's role in renewable energy is becoming increasingly indispensable as nations strive to meet global sustainability goals.

The intermittent nature of renewable energy sources is one of their significant drawbacks. For instance, solar power generation fluctuates with sunlight availability, and wind power depends on wind patterns. AI can mitigate these issues by using predictive analytics to forecast energy production based on weather conditions. These forecasts allow grid operators to plan accordingly, balancing the supply and demand dynamically and reducing reliance on fossil fuels for backup power generation.

AI-enhanced predictive maintenance is another crucial application in renewable energy. Wind turbines and solar panels are subject to wear and tear, which, if not addressed promptly, can lead to significant downtime and loss of energy production. Machine learning algorithms analyze data from sensors embedded in these devices to predict potential failures before they occur. This proactive maintenance approach not only extends the life of the equipment but also ensures continuous and efficient energy production.

In addition to improving the reliability and efficiency of individual renewable energy systems, AI also plays a pivotal role in managing smart grids. Traditional grids were designed for a one-way flow of electricity—from centralized power plants to consumers. However, the rise of decentralized renewable energy installations, like rooftop solar panels, requires a more sophisticated grid management system. AI algorithms help balance the two-way flow of electricity, ensure grid stability, and prevent blackouts by analyzing and reacting to real-time data from various grid components.

Energy storage is a vital element in overcoming the intermittency of renewable energy sources. AI can optimize the use of batteries and other storage technologies by predicting when to store excess energy and when to release it. This intelligent energy management ensures that renewable energy can be reliably available even when production is low, thereby enhancing the overall reliability of renewable energy systems.

Furthermore, AI can assist in the design and placement of renewable energy installations. For example, machine learning models can analyze geographical data, wind patterns, and sunlight exposure to identify the most suitable locations for new wind farms or solar panels. This site optimization leads to higher energy yields and better investment returns.

In the realm of consumer energy management, AI-enabled smart home systems can interact with renewable energy sources to optimize energy use. These systems can schedule high-energy-consuming tasks, like charging electric vehicles or running dishwashers, during periods of high renewable energy production, thereby reducing strain on the grid and lowering electricity costs for consumers.

One can't overlook the role of AI in advancing research and development in renewable energy technologies. By analyzing vast amounts of scientific data, AI can identify patterns and insights that human researchers might miss. This accelerated pace of innovation leads to the development of more efficient and cost-effective renewable energy solutions.

Moreover, the integration of AI and blockchain technology offers promising prospects for peer-to-peer energy trading. AI can manage these decentralized transactions, ensuring transparency and security while enabling consumers to sell excess energy from their personal renewable installations to others on the grid. This democratization of energy trading promotes the adoption of renewable energy and fosters a more sustainable energy ecosystem.

AI's capabilities extend beyond mere optimization and maintenance; it can also play a crucial role in policy-making and regulatory compliance. By providing data-driven insights and simulations, AI helps policymakers understand the long-term implications of their decisions and craft regulations that support the growth of renewable energy. This data-centric approach ensures that policies are both effective and adaptable to changing technological landscapes.

Integration of AI in renewable energy isn't just about improving current technologies; it's also a driver for innovation. AI can simulate and predict performance outcomes for new technologies like advanced photovoltaics or next-generation wind turbines. These simulations save time and resources in the experimental phase, leading to faster development cycles and quicker implementation of groundbreaking technologies.

Looking at the bigger picture, AI provides the analytical firepower to tackle large-scale challenges such as climate change and resource scarcity. Utilizing AI in renewable energy goes beyond operational efficiencies; it serves as a cornerstone for a global strategy aimed at achieving sustainability. AI-driven solutions can provide precise metrics on carbon footprints, energy savings, and the environmental impact of renewable energy projects, enabling stakeholders to make informed decisions and track progress toward sustainability goals.

In summary, the integration of AI into renewable energy systems is not merely an enhancement; it's a revolution. By addressing the inherent challenges such as intermittency, maintenance, and grid management, AI ensures that renewable energy becomes a reliable and efficient cornerstone of modern energy strategies. Its applications span predictive analytics, smart grids, energy storage, and even policy development, demonstrating its versatility and indispensability in this sector.

As we move forward, the synergy between AI and renewable energy promises to unlock unprecedented efficiencies and innovations, setting the stage for a sustainable and resilient energy future. By leveraging AI's capabilities, society can not only accelerate the transition to renewable energy but also ensure that this transformation is both economically viable and environmentally responsible. AI's role in renewable energy encapsulates a future where technology meets sustainability, driving progress toward a cleaner and more prosperous world.

Efficient Resource Management

As the world embraces the rapid integration of Artificial Intelligence (AI) and the Internet of Things (IoT), the energy and utilities sector stands at the forefront of transformation. Efficient resource management is emerging as a critical component in achieving sustainability, reducing operational costs, and enhancing system reliability. By leveraging AIoT, utilities can optimize energy production, distribution, and consumption patterns, resulting in a smarter and more resilient grid.

The heart of efficient resource management lies in predictive analytics. Using advanced AI algorithms, utilities can forecast energy demand and supply with incredible accuracy. These models analyze historical data, weather patterns, and consumption behaviors to predict short-term and long-term energy needs. This insight allows energy providers to balance supply and demand, reducing waste and preventing overproduction. Predictive analytics also helps in anticipating potential system failures, thereby minimizing downtime and improving service reliability.

Moreover, smart meters and IoT sensors play a pivotal role in gathering real-time data. These devices provide granular insights into energy usage across various segments of the grid. They monitor everything from household consumption to industrial energy use, offering a comprehensive view of the entire system's performance. Real-time data collection is invaluable for operations, enabling utilities to make informed decisions promptly. Additionally, real-time monitoring helps in identifying inefficiencies and potential faults, facilitating timely maintenance and repair.

Another critical aspect is demand response. AIoT enables dynamic demand response systems that allow utilities to adjust consumption patterns in response to supply constraints or price signals. For example, during peak demand periods, AI-driven platforms can automatically reduce consumption by temporarily turning off non-essential devices or adjusting thermostat settings in smart homes and commercial buildings. This not only stabilizes the grid but also brings significant cost savings for consumers and providers alike. Such systems exemplify how the interplay between AI and IoT can lead to more efficient and flexible energy management.

Renewable energy sources, such as solar and wind, introduce variability into the grid due to their intermittent nature. AIoT technologies offer a solution by enabling more sophisticated energy storage and distribution systems. AI algorithms can predict renewable energy output based on environmental data and adjust storage and distribution to match predicted demand. IoT-enabled energy storage systems, like smart batteries, can store excess energy generated during peak production times and release it when production dips. This ensures a consistent energy supply and maximizes the use of renewable resources, significantly contributing to sustainability goals.

Enhancing grid resilience is another critical benefit of efficient resource management. By integrating AI and IoT, utilities can build a more adaptable and robust grid. AI can identify patterns and anomalies that precede equipment failure, enabling proactive maintenance and reducing the likelihood of outages. IoT sensors monitor the health and performance of grid infrastructure in real-time, providing continuous feedback and early warnings. This helps in maintaining operational integrity and quickens response times during emergencies, thus preserving service continuity.

Energy conservation is also significantly boosted by the advent of AIoT. Smart home technologies, powered by AI, manage appliances and systems to minimize unnecessary energy use. For instance, smart thermostats learn usage patterns and adjust heating and cooling systems to optimal settings automatically. Similarly, AI-driven lighting systems can turn off lights when rooms are unoccupied and adjust brightness based on natural light availability. These small adjustments accumulate significant energy savings over time and reduce the overall carbon footprint.

Furthermore, AIoT facilitates the integration of distributed energy resources (DERs) into the energy grid. DERs, which include rooftop solar panels, small wind turbines, and community energy storage systems, are becoming increasingly common. AI algorithms manage these diverse and decentralized sources, optimizing their contribution to the grid. IoT devices ensure seamless communication between DERs and central control systems, enhancing coordination and efficiency. This distributed approach not only diversifies energy production but also increases grid reliability and resilience.

Ultimately, efficient resource management through AIoT has profound implications for the overall energy sector strategy. By implementing these advanced technologies, utilities can achieve operational excellence, improve customer satisfaction, and contribute to global sustainability efforts. The future of energy management lies in the symbiotic relationship between intelligent systems and vast networks of connected devices, driving us towards a more resource-efficient world.

In conclusion, the integration of AI and IoT in energy and utilities presents an unparalleled opportunity to revolutionize resource management. By leveraging predictive analytics, real-time data, demand response, renewable integration, and grid resilience strategies, utilities can navigate the complexities of modern energy demands. This transformation fosters a sustainable future, ensuring that energy systems can meet the challenges of tomorrow while delivering reliable and efficient services today.

CHAPTER 10:
TRANSPORTATION AND LOGISTICS

Transportation and logistics are experiencing a seismic shift with the integration of AI and IoT, paving the way for a future where efficiency and safety are paramount. Picture a world where autonomous vehicles reduce traffic congestion, AI-enhanced fleet management optimizes delivery routes, and sophisticated traffic management systems significantly cut down commute times. Sensors and real-time data analysis transform how goods are tracked, ensuring timely deliveries while minimizing costs and environmental impact. As these technologies continue to evolve, the dream of a seamlessly connected transportation network becomes increasingly tangible, promising not just a boost in productivity, but a leap towards a safer, smarter, and more sustainable world.

Autonomous Vehicles

Autonomous vehicles are poised to revolutionize transportation and logistics through their ability to navigate without human intervention. These vehicles leverage a combination of artificial intelligence (AI) and Internet of Things (IoT) technologies, allowing them to perceive their environment, make real-time decisions, and communicate with other devices and infrastructure. They epitomize the transformative potential of AI and IoT integration as they promise to create safer, more efficient, and more sustainable transportation systems.

At the heart of autonomous vehicles is AI, which provides the computational foundation for interpreting data from various sensors, such as cameras, LIDAR, and radar. These sensors generate a continuous stream of data, creating a detailed, multi-dimensional map of the vehicle's surroundings. AI algorithms process this data, enabling the vehicle to recognize objects, understand road conditions, and predict the movement of other road users. This real-time interpretation of complex environments is what enables autonomous vehicles to navigate safely and efficiently.

Equally important is the IoT infrastructure that supports autonomous vehicles. IoT enables vehicles to connect with each other (vehicle-to-vehicle, or V2V), with road infrastructure (vehicle-to-infrastructure, or V2I), and even with pedestrians (vehicle-to-everything, or V2X). This interconnectedness facilitates a robust communication network, allowing vehicles to share crucial information such as traffic congestion, road hazards, and optimal routing. The synergy between AI and IoT not only enhances the vehicle's autonomous capabilities but also contributes to a more integrated and coordinated transportation system.

The impact of autonomous vehicles on logistics is profound. Traditional logistics involve manual driving, which is labor-intensive, prone to human error, and constrained by regulations such as driver working hours. Autonomous vehicles eliminate these limitations by capable of operating 24/7 without fatigue, significantly increasing delivery speed and reliability. AI-driven route optimization further enhances efficiency by determining the most time- and fuel-efficient paths, reducing both operational costs and environmental impact.

Safety, a paramount concern in transportation, stands to benefit enormously from autonomous vehicles. Human errors account for the majority of road accidents, and autonomous vehicles are designed to mitigate these errors by adhering strictly to traffic laws and reacting instantaneously to hazards. Advanced AI systems can predict and avoid potential collisions, even in complex situations, offering a level of safety that human drivers cannot match. Reduced accident rates mean fewer casualties, less property damage, and lower insurance costs, thus contributing to a safer and more secure transport environment.

Sustainability is another key advantage of autonomous vehicles. They are more likely to be electric, reducing the reliance on fossil fuels and cutting greenhouse gas emissions. Autonomous vehicles can also adopt more efficient driving techniques, such as smooth acceleration and braking, which further lowers energy consumption. Moreover, their ability to drive closer together in platoons reduces aerodynamic drag, enhancing fuel efficiency. These factors collectively contribute to a more sustainable and environmentally friendly transportation system.

In urban environments, autonomous vehicles could significantly alleviate traffic congestion. Utilizing AI for real-time traffic management, these vehicles can reroute dynamically to avoid congested areas and optimize the flow of traffic. Smart traffic signals, which adjust in real-time based on vehicle data, can further streamline traffic flow. This integration reduces idle times at intersections and the overall time spent on the road, contributing to less congestion and a smoother, more predictable urban traffic experience.

The implementation of autonomous vehicles isn't without challenges. Technologically, achieving full autonomy requires overcoming obstacles related to sensor accuracy, data processing speed, and software reliability. The diverse and unpredictable nature of real-world driving conditions is a formidable barrier. Weather conditions such as heavy rain or snow can obstruct sensor functions, while unexpected road events require prompt and accurate decision-making.

Ethical and regulatory considerations also need addressing. Autonomous vehicles face moral dilemmas, such as deciding the least harmful action during unavoidable collisions. Additionally, regulations must evolve to accommodate and govern autonomous driving, balancing innovation with public safety. Developing universally accepted standards and legislation is crucial for the widescale adoption of autonomous vehicles.

In shaping the future workforce, autonomous vehicles will likely alter the landscape of employment in transportation and logistics. While they could displace some driving jobs, they will also create new roles focused on vehicle maintenance, IT support, and fleet management. Upskilling the current workforce to adapt to these new demands will be essential. Educational programs and training initiatives will play a vital role in preparing workers for the technological shifts brought about by autonomous vehicles.

Lastly, autonomous vehicles promise to enhance the accessibility of transportation. They can provide mobility solutions for individuals who are unable to drive, such as the elderly or disabled. Autonomous ride-sharing services can offer more affordable and accessible transport options, particularly in underserved areas where traditional public transport isn't viable. This inclusivity ensures that the benefits of autonomous vehicles extend to a broader segment of the population, promoting social equity in mobility.

In conclusion, autonomous vehicles represent a significant leap forward in the integration of AI and IoT within the transportation and logistics sector. They offer compelling benefits in terms of safety, efficiency, sustainability, and accessibility. While challenges remain, the continued advancement of technology, coupled with thoughtful regulatory frameworks and workforce adaptation strategies, will pave the way for the successful adoption of autonomous vehicles. This transformative technology holds the promise of reshaping our transportation systems for the better, making them smarter, greener, and more inclusive.

Fleet Management

In the realm of transportation and logistics, fleet management stands as a pivotal component, significantly impacted by the integration of Artificial Intelligence (AI) and the Internet of Things (IoT). This transformation isn't just about using new technologies; it's creating a more efficient, safe, and cost-effective way to manage vehicle fleets, large or small.

Fleet management has traditionally involved tracking vehicles, managing maintenance schedules, and ensuring that drivers adhere to regulations. However, with AI and IoT, there's now a new layer of intelligence and real-time responsiveness. Fleet managers can now monitor their vehicles' every move, diagnose issues before they become critical, and even predict future maintenance needs.

One of the most transformative aspects of AI in fleet management comes from predictive analytics. AI algorithms analyze vast amounts of data from various sources, including vehicle telematics, driver behavior, and environmental conditions. This analysis allows fleet managers to predict potential issues and schedule maintenance proactively, thus reducing downtime and avoiding costly repairs. Instead of reacting to breakdowns, companies can now foresee and mitigate them.

IoT devices play a critical role in this ecosystem. Sensors and connected devices installed in vehicles continuously collect and transmit data. These devices monitor a multitude of parameters such as engine health, fuel consumption, tire pressure, and driver behavior. This data is then sent to a central management system where AI algorithms process it to provide actionable insights.

Consider the example of real-time vehicle tracking. IoT-enabled GPS systems provide continuous location data, allowing fleet managers to optimize routes in real-time. If a certain route is congested, the system can suggest an alternative path, saving time and reducing fuel consumption. Additionally, geofencing capabilities allow for improved security measures, alerting managers if a vehicle deviates from its pre-determined route.

Improved driver safety is another significant benefit of AI and IoT integration. Advanced driver assistance systems (ADAS) leverage AI to monitor and assist driving. Features such as lane departure warnings, collision avoidance systems, and fatigue detection are all driven by AI, potentially saving lives and reducing accident-related costs. Moreover, fleet managers can monitor driver behavior in real-time and provide feedback to ensure safer driving practices.

Furthermore, fleet management systems powered by AI enhance fuel efficiency. By analyzing data on driving patterns, traffic conditions, and vehicle performance, AI can suggest optimal driving speeds, routes, and even idling times. This leads to a significant reduction in fuel consumption and greenhouse gas emissions, contributing to both cost savings and environmental benefits.

Inventory management within fleet operations has also seen improvements thanks to the integration of AI and IoT. AI can track and manage the inventory of spare parts, ensuring that the necessary components are available when needed. This minimizes downtime and streamlines the maintenance process.

Additionally, fleet managers can now utilize AI-driven dashboards that offer comprehensive views of fleet operations. These dashboards present real-time data visualizations and predictive analytics, enabling informed decision-making. Managers can assess vehicle utilization rates, optimize asset allocation, and ensure that the fleet is operating at peak efficiency.

Compliance and regulatory adherence have always been challenging areas for fleet management. However, AI and IoT facilitate better compliance with safety and emissions standards. Automated logging of driving hours, real-time monitoring of emissions data, and AI-driven documentation processes ensure that fleets meet regulatory requirements effortlessly.

Insurance is another sector that benefits from AI and IoT in fleet management. By leveraging data from connected vehicles, insurance companies can offer usage-based insurance policies. These policies reflect real driving patterns, potentially lowering premiums for safe drivers and encouraging better driving habits. Additionally, in the event of an accident, AI-powered systems can analyze telematics data to determine the cause and liability, speeding up the claims process.

The financial implications of AI and IoT in fleet management are profound. Initial investments in these technologies might seem substantial, but their long-term savings make them worthwhile. Reduced maintenance costs, optimized fuel consumption, improved asset utilization, and lower insurance premiums all contribute to significant cost reductions.

Moreover, customer satisfaction improves as well. Timely deliveries, fewer delays, and reliable service bolster a company's reputation. AI can also aid in customer communication, providing real-time updates on shipment status, expected arrival times, and potential delays.

As AI and IoT continue to evolve, so will their applications in fleet management. Future advancements may include autonomous vehicles, which could revolutionize the industry by reducing labor costs and further increasing efficiency. AI-driven predictive maintenance will become even more accurate, driven by more extensive data sets and improved algorithms.

The integration of Blockchain with AI and IoT also holds promise for enhancing fleet management. Blockchain can ensure data integrity and transparency in supply chains, further improving trust and efficiency. Smart contracts enabled by Blockchain can automate various aspects of fleet management, including payments and compliance verification.

While the benefits are numerous, it's essential to recognize the challenges associated with this technological transformation. Data security and privacy concerns are paramount, given the vast amount of sensitive information generated by connected vehicles. Robust cybersecurity measures must be in place to protect against data breaches and cyberattacks.

The transition to AI and IoT-enabled fleet management requires a shift in skill sets for fleet managers and technicians. Continuous training and education are necessary to keep up with rapidly evolving technologies. Collaboration with technology providers and ongoing support are crucial to effectively implement and maintain these systems.

In conclusion, the combination of AI and IoT has brought about a paradigm shift in fleet management. This transformation enhances operational efficiency, improves safety, reduces costs, and contributes to environmental sustainability. The journey ahead promises even more remarkable advancements, making fleet management smarter and more responsive to the dynamic demands of the transportation and logistics industry.

AI in Traffic Management

In the bustling realm of modern cities, efficient traffic management stands out as a critical challenge. Urban populations are growing rapidly, and with this growth comes the increased strain on existing traffic infrastructure. Enter AI: a transformative force that holds the promise of revolutionizing how we manage and optimize traffic flow. By integrating AI into traffic management systems, city planners are moving towards creating smarter, safer, and more efficient urban environments.

One of AI's most notable contributions to traffic management lies in its ability to process and analyze vast amounts of data in real-time. Traditional traffic systems, which often rely on static signals and manual adjustments, struggle to adapt to the dynamic nature of urban traffic. AI-driven systems, on the other hand, use data from various sources—including cameras, sensors, and GPS devices—to monitor traffic patterns continuously and make real-time adjustments to signal timings and route suggestions. This fluid adaptability minimizes congestion and reduces travel times, leading to less frustration for drivers and more efficient use of roadways.

Another significant advantage of AI in traffic management is its predictive capabilities. By leveraging historical data and real-time information, AI algorithms can forecast traffic conditions and potential bottlenecks before they occur. For instance, during events like concerts or sports games, AI systems can predict surges in traffic and preemptively adjust traffic lights and suggest alternate routes to mitigate congestion. This proactive approach stands in stark contrast to the reactive nature of traditional traffic management methods.

Moreover, AI-driven traffic management systems enhance road safety by identifying and addressing potential hazards. Advanced techniques such as computer vision and machine learning can detect abnormal driving behaviors, stalled vehicles, or accidents in real-time. When these incidents are identified, AI can trigger immediate responses, from alerting authorities to dynamically rerouting traffic and dispatching emergency services. Such rapid interventions not only save lives but also ensure that traffic flow resumes as quickly as possible following disruptions.

Public transportation systems also benefit significantly from AI integration. AI can optimize bus and train schedules based on real-time passenger data, improving the efficiency and reliability of public transit. Predictive analytics can determine peak usage times and suggest the best allocation of resources, ultimately enhancing the overall commuter experience and encouraging more people to use public transport, thus reducing the number of individual vehicles on the road.

Integrating AI into traffic management also paves the way for the implementation of smart intersections. These intersections, equipped with AI-powered sensors and communication devices, can interact with vehicles and pedestrians to optimize traffic flow and enhance safety. For instance, smart intersections can prioritize emergency vehicles by adjusting traffic lights to create green corridors, ensuring these vehicles reach their destinations swiftly and safely. Similarly, they can detect pedestrians and cyclists, giving them more time to cross streets safely.

The environmental impact of AI in traffic management is another crucial aspect. Traffic congestion is a significant contributor to urban air pollution, producing higher emissions from idling vehicles. By improving traffic flow and reducing congestion, AI systems can significantly cut down on vehicle emissions, contributing to cleaner, healthier urban environments. Furthermore, optimized traffic systems result in better fuel efficiency, reducing the overall carbon footprint of city transportation networks.

In addition to ground transportation, AI is making strides in managing other forms of urban mobility, such as e-scooters and bike-sharing programs. These new modes of transport are becoming increasingly popular in cities worldwide, creating additional layers of complexity in traffic management. AI's capability to analyze usage patterns and predict demands ensures that bikes and e-scooters are available where and when they're needed, balancing supply and demand efficiently and integrating seamlessly into the overall traffic system.

The collaboration between AI and IoT further enhances the potential of traffic management systems. IoT devices embedded in infrastructure and vehicles collect data that AI algorithms then analyze to provide actionable insights. This synergy allows for a highly interconnected traffic ecosystem where every component communicates and collaborates to create a cohesive and efficient network. For example, IoT-enabled vehicles can share real-time information about their speed and direction with each other and with traffic management centers, enabling smoother traffic flow and reducing the likelihood of accidents.

However, the widespread adoption of AI in traffic management is not without challenges. Privacy concerns and data security are significant issues that must be addressed to ensure public trust and acceptance. The vast amounts of data collected by AI systems include sensitive information about individuals' movements and behaviors. Robust measures must be implemented to protect this data from unauthorized access and misuse.

Interoperability and standardization are other critical challenges. Traffic management systems often include components from various manufacturers and service providers. Ensuring that these disparate systems can effectively communicate and work together requires the establishment of common standards and protocols. This interoperability is essential for creating a seamless and efficient traffic management ecosystem.

Despite these challenges, the potential benefits of AI in traffic management are immense. As technology continues to advance, AI systems will become even more sophisticated and capable, offering new solutions to urban traffic woes. The integration of AI not only promises to transform our current traffic systems but also paves the way for future advancements in autonomous vehicles and smart city infrastructure.

In conclusion, AI's role in traffic management represents a significant leap forward in our quest for smarter, safer, and more efficient urban environments. By harnessing the power of real-time data analysis, predictive capabilities, and adaptive systems, AI has the potential to revolutionize how we navigate our cities and manage traffic flow. While challenges such as privacy, security, and interoperability remain, the continued development and integration of AI in traffic management hold great promise for the future of urban mobility.

CHAPTER 11:
FINANCE AND BANKING

As AI and IoT converge to reshape the finance and banking sectors, the impact is both profound and far-reaching. Traditional banking processes are becoming increasingly automated, leading to faster and more secure transactions. AI-driven algorithms are enhancing fraud detection capabilities, sifting through vast amounts of data to identify anomalous activities in real time. Moreover, customer service is being revolutionized by AI-powered chatbots and virtual assistants, offering personalized support around the clock. IoT devices are introducing new dimensions to financial transactions, enabling seamless and secure mobile payments and expanding access to financial services in underserved regions. This confluence of AI and IoT is not only optimizing operational efficiencies but also fostering greater financial inclusion and trust in the digital economy, paving the way for a future where financial interactions are as intuitive as they are secure.

Fraud Detection

In the finance and banking sector, fraud has always been a significant challenge, causing financial institutions to lose billions of dollars annually. Traditional methods of fraud detection were largely rule-based and reactive in nature, often catching fraud after the fact. However, with the advent of AI and IoT, the landscape of fraud detection is experiencing a revolutionary change. By leveraging advanced algorithms, machine learning, and real-time data from IoT devices, banks can now detect and prevent fraud with unprecedented accuracy and speed.

AI's ability to analyze vast amounts of data in real-time is particularly useful in identifying fraud patterns that might elude human analysts. Machine learning models can be trained to recognize unusual activities, such as atypical transaction amounts, unusual account access patterns, or strange geographical locations. Once these anomalies are flagged, the system can trigger alerts for further investigation. This transition from rule-based systems to behavior-based models marks a significant leap forward in fraud detection capabilities.

For instance, imagine a scenario in which a user typically makes purchases in their hometown. If a transaction is suddenly initiated from a foreign country that this user has never visited, the system can quickly identify this as a potential risk and halt the transaction, pending verification. This real-time intervention capability, driven by AI, helps in mitigating fraud losses significantly. Moreover, as the system continues to learn and adapt, its accuracy improves, resulting in fewer false positives and more effective fraud prevention.

IoT devices contribute significantly to the overall fraud detection framework. Wearables, smart ATMs, and connected banking apps generate massive amounts of data, which, when analyzed using AI algorithms, can provide deep insights into user behavior. By creating a holistic picture of user activity across multiple devices and platforms, it becomes easier to detect inconsistencies that could indicate fraudulent activities.

One of the primary benefits of incorporating AI and IoT in fraud detection is the ability to analyze data from multiple sources in real-time. For example, IoT sensors installed at ATMs can monitor physical interactions, such as card insertions and withdrawals, while AI algorithms analyze transaction data. By correlating these data points, the system can detect if a card has been cloned or if a skimming device has been attached to the ATM.

The integration of AI also enhances the ability to combat identity theft, a growing concern in the digital age. Advanced biometric authentication methods, such as facial recognition and fingerprint scanning, are being employed to ensure that the person making the transaction is indeed the legitimate account holder. These systems use AI to compare live data with stored biometric templates in milliseconds, adding an extra layer of security that is difficult for fraudsters to bypass.

Another critical application of AI in fraud detection is in analyzing transactional patterns within large datasets. Traditional systems might flag a single irregular transaction, but AI can look at sequences of transactions to discern complex patterns that indicate fraud. For example, it can identify cases where small amounts of money are moved across multiple accounts—an activity often associated with money laundering. Additionally, machine learning models are capable of evolving over time, incorporating new data to refine their algorithms and improve detection rates, which is particularly beneficial in dealing with evolving fraud tactics.

Aside from transactional data, AI and IoT can also use environmental data to detect potential fraud. IoT devices can monitor the environment where transactions take place, providing contextual information that can be vital for fraud detection. For instance, AI can analyze factors like time of day, location data, and even the type of device used for the transaction. By combining this environmental data with transactional data, the system can make more informed decisions about the legitimacy of a transaction.

User behavior analysis is another domain where AI's capabilities shine. By creating detailed profiles of how individual users typically interact with their accounts, AI can discern deviations with high precision. For instance, if a user who usually withdraws money weekly suddenly starts making large, frequent transfers, AI can flag this behavior for review. Furthermore, AI systems can cross-reference this behavior with industry-wide patterns to ensure that the deviation isn't merely a new trend but an actual risk indicator.

While the benefits of integrating AI and IoT in fraud detection are immense, it's crucial to address the challenges and potential risks. For one, the reliance on vast amounts of data necessitates stringent data security measures to protect sensitive information from breaches. Additionally, there's the risk of over-reliance on these technologies, which might lead organizations to overlook the importance of human oversight. Combining human intuition with AI's analytical power creates a balanced approach, ensuring robust fraud detection mechanisms.

Moreover, the use of AI in fraud detection must be approached with transparency and fairness. As AI systems make decisions that can significantly impact individuals, it's essential to ensure these systems are free from biases that could lead to unfair treatment. For example, a biased AI model might flag transactions from specific regions as more likely to be fraudulent, leading to unnecessary inconveniences for legitimate users. Therefore, constant monitoring, evaluation, and updating of AI models are required to maintain fairness and accuracy.

The future of fraud detection in finance and banking is undoubtedly promising, with AI and IoT at the helm. As these technologies continue to evolve, they will enable even more sophisticated and accurate fraud detection mechanisms. Financial institutions must stay abreast of these developments and continually adapt their systems to harness the full potential of AI and IoT in safeguarding against fraud. By doing so, they can offer their customers a more secure banking experience while minimizing financial losses.

In conclusion, the integration of AI and IoT in fraud detection within the finance and banking sector represents a significant advancement over traditional methods. These technologies enable real-time analysis, pattern recognition, and user behavior profiling, providing robust mechanisms to detect and prevent fraudulent activities. While challenges exist, particularly regarding data security and bias, the benefits far outweigh the drawbacks. As AI and IoT technologies continue to advance, their role in fraud detection will only become more vital, ushering in a new era of secure and efficient financial transactions.

AI-Powered Customer Service

The financial services sector, particularly banking, has historically been seen as a conservative industry. However, the integration of AI-driven customer service solutions has radically shifted this perception. By leveraging AI, banks and financial institutions are providing more personalized, efficient, and 24/7 customer service, transforming how they interact with their clientele.

AI-powered chatbots are one of the most visible applications in this transformation. These chatbots use natural language processing (NLP) and machine learning algorithms to understand and respond to customer queries. What sets them apart is their ability to learn and improve from each interaction, offering increasingly accurate and relevant responses over time. Not only do they handle routine inquiries—such as balance inquiries, transaction history, and basic troubleshooting—but they can also escalate more complex issues to human agents, ensuring seamless service continuity.

Moreover, AI-driven customer service systems can personalize interactions with customers by analyzing vast amounts of data. By scrutinizing transaction histories, spending patterns, and financial behaviors, AI can offer tailored financial advice, product recommendations, and customized solutions. This level of personalization was previously unachievable with traditional customer service models, which heavily relied on generic, one-size-fits-all advice and support.

Additionally, advanced voice recognition technologies enable AI systems to understand and process verbal requests with remarkable accuracy. These systems are integrated into phone services, allowing customers to resolve their concerns without waiting for a human agent. As voice assistants become more sophisticated, they can handle increasingly complex inquiries, from intricate product details to specific transaction queries.

The implementation of AI in customer service doesn't just benefit the customers; it also significantly impacts the operational efficiency of financial institutions. AI systems can manage a high volume of inquiries simultaneously, reducing wait times and operational costs. Financial institutions can allocate their human resources to more complex, high-value tasks such as financial planning and advisory services. This leads to higher job satisfaction among employees and a more streamlined, effective customer service operation.

Another crucial aspect of AI-powered customer service in banking is its role in fraud detection and prevention. AI systems can monitor transactions in real-time, identifying suspicious activities based on established patterns and anomalies. When unusual behavior is detected, the system can instantly alert both the customer and the institution, often before any significant damage is done. This proactive approach not only enhances security but also instills confidence in customers, knowing that their financial assets are continually monitored.

Boosting customer loyalty and satisfaction is another significant outcome of AI-powered customer service. Positive interactions foster trust and reliability, which are paramount in banking relationships. AI systems' ability to resolve issues quickly and efficiently means customers are less likely to encounter frustrations that could drive them to seek services elsewhere.

Furthermore, AI can be utilized to gather customer feedback and sentiment analysis. By analyzing the language and tone used in customer interactions, AI can identify underlying issues and areas for improvement. This continuous feedback loop helps financial institutions refine their services and better align with customer expectations. Identifying pain points allows banks to address them proactively, creating a more efficient and satisfying customer experience.

One compelling example of AI-driven customer service in banking is the use of robo-advisors. These AI-powered advisory platforms offer investment advice, portfolio management, and financial planning services traditionally provided by human advisors. By leveraging algorithms that take into account market trends, risk tolerance, and individual financial goals, robo-advisors offer tailored financial strategies that are both cost-effective and highly efficient.

The integration of predictive analytics also offers transformative benefits. By predicting future customer needs based on historical data, AI systems can provide timely and relevant advice. For instance, if a customer's spending pattern suggests an upcoming large purchase, the AI system can offer loan options or special saving plans. This predictive capability not only enhances customer satisfaction but also opens new business opportunities for financial institutions.

Data privacy and security remain paramount as financial institutions integrate more AI-driven solutions. There must be strict compliance with regulatory standards to protect sensitive customer information. Robust encryption, regular audits, and adherence to data protection laws ensure that AI systems do not compromise customer trust. Financial institutions must balance innovation with the fundamental need for security, transparency, and ethical data use.

Given these advances, the future of AI in customer service for finance and banking looks highly promising. As algorithms continue to evolve, the depth and quality of AI interactions will only improve. This ongoing evolution is set to make financial services more accessible, transparent, and equitable for all customers, regardless of their technological proficiency.

The fusion of human ingenuity with the relentless efficiency of AI creates a new paradigm in financial services. AI-powered customer service not only meets but often exceeds customer expectations, setting new standards for what exceptional service entails. It's not just about answering questions quickly—it's about anticipating needs, providing insightful advice, and fostering a relationship built on trust and responsiveness.

In the grand tapestry of technological advancement, AI-powered customer service in the finance and banking sector is a vibrant thread, weaving together the possibilities of today with the innovations of tomorrow. This amalgamation of data, technology, and human-centered design ensures that financial institutions remain resilient, adaptive, and profoundly customer-focused.

Financial institutions must continue investing in AI technologies while nurturing the human touch that defines exceptional customer experiences. By doing so, they can navigate the complexities of the digital age and emerge as leaders in a rapidly evolving landscape. The journey of integrating AI in customer service is ongoing, and its potential is vast, promising a future where financial services are smarter, more intuitive, and deeply attuned to the needs of every individual.

In conclusion, AI-powered customer service is not merely a technological upgrade; it's a fundamental shift in how financial institutions engage with their customers. As AI technologies grow more sophisticated, the benefits will ripple through every aspect of banking, heralding a new era of customer-centric financial services. This transformation underscores the profound impact of AI on our daily lives and highlights the boundless possibilities for innovation in the finance sector.

IoT in Financial Transactions

Finance and banking are traditionally conservative sectors, but the integration of IoT is significantly transforming how transactions are conducted. The ability of IoT devices to collect, transmit, and analyze data in real-time is providing new opportunities for enhancing the efficiency, security, and convenience of financial transactions. From contactless payments to smart ATMs, IoT is reshaping the financial landscape.

One of the most noticeable impacts of IoT in financial transactions is the rise of contactless payments. Consumers no longer need to carry cash or even cards; instead, they can use their smartphones, smartwatches, or other wearables equipped with NFC (Near Field Communication) technology to make payments. This seamless transaction method not only speeds up the payment process but also reduces physical contact, which has become crucial in the wake of global health concerns.

Furthermore, IoT devices can enhance the security of financial transactions through better authentication mechanisms. For instance, biometric sensors integrated into IoT devices can verify a person's identity through fingerprints, facial recognition, or even voice recognition. These advanced authentication techniques make it increasingly difficult for unauthorized individuals to gain access to financial accounts, thereby reducing fraud and enhancing consumer trust.

In addition to security, IoT is paving the way for more personalized banking experiences. Smart banking applications can use data gathered from IoT devices to offer tailored financial advice and services. For example, a smart financial app might analyze your spending habits and send you alerts or suggestions on budgeting more effectively. Such personalized services are possible because IoT devices constantly collect and transmit data, providing banks with real-time insights into their customers' behaviors and needs.

Charlie Morgan

The implementation of IoT in ATMs is another area where significant advancements are being observed. Smart ATMs can now offer more than just cash withdrawals. They can provide a range of services, including instant loan approvals and cryptocurrency transactions, thanks to their connectivity and ability to process data quickly. Moreover, these smart machines can use sensors to detect and prevent tampering attempts, enhancing overall security.

IoT also plays a crucial role in improving the efficiency of financial institutions. For example, banks can monitor the status of their equipment and facilities using IoT sensors, enabling predictive maintenance to avoid downtime. Similarly, IoT can streamline the management of cash flow within ATMs and branches, using real-time data to predict cash demand and optimize cash distribution. Such efficiencies not only save costs but also improve customer satisfaction by ensuring services are always available.

A revolutionary aspect of IoT in financial transactions is the integration with blockchain technology. Combining IoT devices with blockchain can provide enhanced transparency and traceability for financial transactions. For instance, IoT-enabled supply chain financial transactions can be automatically recorded in a blockchain ledger, ensuring each step is tracked and verified. This hybrid approach can significantly reduce fraud and errors, providing a reliable audit trail.

Moreover, IoT and AI together can provide powerful tools for risk management and fraud detection. Advanced algorithms can process the immense amount of data collected by IoT devices to identify unusual patterns that might indicate fraudulent activities. Real-time analysis can trigger alerts and even automatically block suspicious transactions, thereby preventing potential financial losses before they occur.

As IoT devices become more ubiquitous, ethical and privacy considerations come to the forefront. Financial institutions must ensure robust data protection measures to safeguard the vast amounts of personal and financial data being transmitted and stored. This involves using encryption, secure communication channels, and complying with regulations such as GDPR to protect consumer privacy.

Another area where IoT can make a substantial impact is in financial inclusion. In regions with limited access to traditional banking infrastructure, IoT devices such as mobile phones can provide crucial banking services to underserved populations. Microloans, savings accounts, and payment services can be accessed via mobile technologies, thus bringing financial services to remote and rural areas, fostering economic growth and stability.

The role of IoT in financial transactions extends to investment management as well. IoT devices can provide real-time data on market conditions, asset performance, and even geopolitical events that might impact investments. Automated trading platforms can use this data to make informed decisions, executing trades at optimal times with minimal human intervention. This not only improves investment accuracy but also enables investors to respond swiftly to market changes.

In conclusion, the rapid integration of IoT in financial transactions marks a significant evolution in the finance and banking sector. By enhancing security, offering personalized services, and improving operational efficiency, IoT is setting new standards for financial transactions. As we continue to explore the potential of IoT, it's clear that its impact on finance and banking will be profound and far-reaching, driving innovation and transforming how we manage and interact with money. The journey is just beginning, and the possibilities are as vast as they are exciting.

CHAPTER 12:
REAL-WORLD AIoT APPLICATIONS

In this chapter, we'll delve into the nitty-gritty of how AIoT—or the harmonious integration of Artificial Intelligence and the Internet of Things—has transcended theoretical frameworks to revolutionize real-world applications. Whether it's in transforming urban environments into smart cities, optimizing healthcare through predictive diagnostics, or enhancing manufacturing with intelligent automation, AIoT's practical implications are vast and varied. We'll explore compelling case studies and success stories that underscore the transformative power of AIoT, illuminating lessons learned from both triumphs and challenges. Prepare to be inspired by how these technologies are not just improving efficiencies but also driving innovation across industries, setting the stage for a future replete with limitless possibilities.

Case Studies

The integration of Artificial Intelligence (AI) and the Internet of Things (IoT), known as AIoT, is transforming how industries operate and individuals interact with technology. This section delves into several compelling case studies that illustrate the practical and impactful applications of AIoT in various sectors. These examples not only highlight the technological advancements but also show the tangible benefits and challenges faced during implementation.

One of the most compelling case studies comes from smart homes, where tech companies have developed sophisticated AIoT systems that automate and optimize household tasks. For instance, consider the system implemented by Nest, a Google-owned company. Nest uses machine learning algorithms to learn the daily routines of a household and adjusts the thermostat accordingly, resulting in significant energy savings. Users report not only a reduction in their utility bills but also an increased sense of comfort and convenience. Their homes automatically adjust to their preferences without requiring manual inputs. This demonstrates how AIoT can make daily routines more efficient and eco-friendly.

In the healthcare sector, AIoT is revolutionizing patient care through remote monitoring systems. Consider the example of Medtronic, a global leader in medical technology. Medtronic has developed an IoT-enabled insulin pump that continuously monitors blood glucose levels and automatically administers insulin. The AI algorithms within the system predict glucose trends and adjust insulin delivery in real-time. This not only enhances the quality of life for diabetes patients but also reduces the risk of emergencies. A study revealed that patients using this system experienced a significant decrease in hypoglycemic events. By leveraging AIoT, Medtronic has set a new standard for chronic disease management.

The benefits of AIoT are also evident in industrial applications. Take General Electric (GE) as an example. GE has integrated AI and IoT in its manufacturing processes to enable predictive maintenance. By deploying IoT sensors on machinery and using AI to analyze the data, GE can predict when equipment is likely to fail and schedule maintenance before breakdowns occur. This approach has drastically reduced downtime and maintenance costs. In pilot projects, GE reported a 30% reduction in unplanned downtime, showcasing how AIoT can enhance productivity and reduce operational costs in manufacturing.

Another remarkable case study is seen in the agricultural sector with the use of precision farming techniques. John Deere, a well-known brand in agricultural machinery, has adopted AIoT technologies to assist farmers in crop management. Their AIoT-enabled tractors and equipment use sensors to gather data on soil conditions, weather forecasts, and crop health. This data is analyzed by AI algorithms to provide farmers with actionable insights and recommendations. The farmers can then make informed decisions about irrigation, fertilization, and pest control. This has led to higher crop yields and optimized resource usage, as demonstrated in various pilot programs across different regions.

Retail is another area where AIoT is making a significant impact. Amazon Go stores are a perfect example of how AIoT can transform the shopping experience. These stores use a combination of AI, computer vision, and IoT sensors to create a checkout-free shopping experience. Shoppers simply walk in, pick up the items they want, and walk out. The items are automatically detected, and the cost is charged to their Amazon account. This technology not only ensures a seamless and quick shopping experience but also reduces the need for checkout staff, leading to operational efficiencies. The success of Amazon Go stores highlights the potential for AIoT to revolutionize retail operations.

In the realm of smart cities, the city of Barcelona has been a frontrunner in adopting AIoT solutions for urban management. The city's smart lighting system uses IoT sensors to monitor real-time data, such as foot traffic, and adjusts streetlight intensity accordingly. This system has not only resulted in significant energy savings but also improved public safety by ensuring that areas with higher foot traffic are well-lit. Additionally, Barcelona's smart waste management system uses IoT sensors in trash bins to monitor the fill level and route garbage trucks efficiently. These initiatives have led to a cleaner city and reduced waste collection costs, serving as an exemplary model for urban AIoT applications.

Transportation systems are also benefiting from AIoT tech-nologies. The Pittsburgh Traffic21 program is a standout case study. By installing IoT-enabled traffic signals equipped with AI algorithms throughout the city, Pittsburgh has significantly improved traffic flow and reduced congestion. The smart traffic lights use real-time data to adapt their signal phases dynamically, addressing traffic bottlenecks as they develop. This system has led to a 25% reduction in travel time during peak hours, showcasing how AIoT can optimize urban mobility.

One can't discuss transformative AIoT applications without men-tioning Tesla's autonomous vehicles. Tesla has integrated AI and IoT to create a network of self-driving cars that continuously learn and im-prove through data collected from Tesla vehicles on the road. The AI algorithms analyze this data to enhance driving algorithms, making Tesla cars increasingly safer and more efficient. Real-world data has shown that Tesla's AIoT-powered vehicles have far fewer accidents compared to traditional vehicles, providing a glimpse into the future of automotive technology.

Energy management is another sector witnessing significant im-provements through AIoT. The case of Duke Energy's smart grid ini-tiatives is worth noting. Duke Energy has implemented advanced me-tering infrastructure (AMI) and deployed IoT sensors across its grid. These sensors provide real-time data on energy consumption and grid performance, which is then analyzed by AI-predictive models to opti-mize energy distribution. The smart grid has resulted in better load management and reduced energy wastage, resulting in cost savings for both the company and consumers. The enhanced reliability of energy supply has also improved customer satisfaction.

Lastly, a unique application of AIoT is observed in the finance sector with JPMorgan Chase's fraud detection systems. The bank uses AI algorithms integrated with IoT-connected devices to monitor and analyze real-time transactional data. The AI models can identify unusual patterns that may indicate fraudulent activities, alerting bank officials for further investigation. This has resulted in a significant reduction in fraud-related losses and improved the overall security of financial transactions. The case study of JPMorgan Chase exemplifies how AIoT can provide robust solutions for complex problems in the finance industry.

These varied case studies highlight the diverse and impactful ways in which AIoT is being implemented across different sectors. From enhancing daily life in smart homes to optimizing industrial processes, and improving urban management, the potential of AIoT is vast. Each example underscores not only the technological capabilities but also the significant benefits such as cost savings, improved efficiency, and better user experiences. As AIoT continues to evolve, its applications are expected to expand further, bringing transformative changes across even more domains.

These real-world case studies serve as a testament to the profound impact of integrating AI and IoT. They provide concrete evidence of how technology can improve lives, drive efficiency, and foster innovation. As we move forward, these examples offer valuable insights into the endless possibilities and set the groundwork for future advancements in AIoT.

Success Stories

In the vast landscape of AI and IoT's interconnected world, numerous success stories stand as a testament to the revolutionary impact these technologies have had across various industries. These real-world applications not only showcase the versatility and potential of AIoT but also serve as beacons of innovation, guiding future developments.

One spectacular success story is the transformation witnessed within the healthcare sector. Hospitals and clinics worldwide have integrated AIoT to enhance patient care and operational efficiency. For instance, the Mayo Clinic has utilized AI-powered platforms combined with IoT devices to monitor patient vitals continuously. This integration allows for real-time data analytics, significantly improving response times during emergencies and enabling personalized treatment plans. The ability to predict patient deterioration through continuous monitoring has been a game-changer, drastically reducing mortality rates and improving overall patient outcomes.

The agricultural industry has also experienced a profound shift, thanks to AIoT technologies. Outdoor farms equipped with IoT sensors and AI-driven analytics frameworks are now able to monitor soil moisture levels, crop health, and weather conditions with unprecedented precision. A standout example is John Deere's adoption of precision farming technologies. By leveraging AIoT, farmers can now optimize planting schedules, irrigation, and pesticide use, resulting in higher yields and more sustainable farming practices. This move towards smart farming has not only boosted productivity but also alleviated some of the environmental impacts of traditional farming methods.

In the urban sphere, smart cities epitomize the successful fusion of AI and IoT. Cities like Barcelona have implemented an extensive network of IoT sensors to manage urban infrastructure effectively. The AI algorithms process vast amounts of data to optimize traffic flow, reduce energy consumption, and improve waste management. These systems have led to remarkable improvements in air quality, reduced traffic congestion, and made the city more livable for its residents. By leveraging AIoT, Barcelona has set a precedent for other cities looking to harness technology for sustainable urban living.

The energy sector has not been left behind in this technological revolution. Companies like Siemens have introduced smart grid solutions that utilize AI and IoT for efficient energy management. By accurately predicting energy consumption patterns and optimizing distribution, these intelligent grids reduce energy wastage and enhance the reliability of power supply. Notably, Siemens' smart grids have been instrumental in integrating renewable energy sources into the mainstream power supply, aligning with global sustainability goals.

The retail industry has also unlocked significant value from AIoT integration. One remarkable success story is Walmart's utilization of AI and IoT for enhancing supply chain logistics and inventory management. With real-time tracking of inventory levels and advanced data analytics, Walmart can anticipate demand fluctuations, reducing overstock and stockouts. This not only ensures optimal inventory levels but also elevates the shopping experience for customers by ensuring product availability. Moreover, personalized shopping experiences driven by AI algorithms analyzing customer behavior data have further cemented Walmart's leadership in the retail space.

Manufacturing processes have seen a monumental shift with the introduction of AIoT in industrial settings. General Electric (GE) has pioneered the use of predictive maintenance in their manufacturing plants. By deploying IoT sensors on machinery and leveraging AI for real-time data analysis, GE can foresee equipment failures and schedule maintenance before breakdowns occur. This proactive approach has minimized downtime, significantly cut maintenance costs, and maximized production efficiency.

Transportation systems globally have benefited immensely from AIoT advancements. The development of autonomous vehicles by companies like Tesla is a prime example of how AIoT is driving the future of mobility. These vehicles utilize an amalgamation of sensors, GPS, and advanced AI algorithms to navigate roads autonomously and safely. The implementation of this technology promises to revolutionize not just personal transportation but also logistics and supply chain operations, offering safer and more efficient transit solutions.

In the finance sector, JP Morgan Chase has successfully employed AIoT for fraud detection and risk management. By using IoT to track transaction patterns and AI to analyze this data intricately, the bank can detect fraudulent activities in real-time, safeguarding assets and maintaining customer trust. Their intelligent systems are also able to provide personalized financial advice to clients, streamlining customer service and enhancing user experience.

Retail giant Amazon showcases another brilliant AIoT success story through its cashier-less Amazon Go stores. These stores utilize a network of cameras and sensors to track items customers pick up and put back, which is then analyzed by AI to automatically charge their accounts upon exit. This seamless shopping experience has not only reshaped consumer expectations but also highlighted the potential for similar applications in various retail contexts.

The precision and responsiveness brought by AI and IoT to emergency services cannot be overstated. Consider the smart fire detection systems implemented in San Francisco. These systems use IoT sensors to monitor environmental conditions constantly and AI to predict fire outbreaks with high accuracy. Rapid alerts ensure that emergency responders can act swiftly, significantly mitigating damage and saving lives.

Education has tapped into AIoT's potential to create smarter learning environments. Schools equipped with smart classroom technologies like those in Singapore have seen a transformation in teaching and learning. IoT devices paired with AI tools assist in delivering personalized learning experiences, tracking student performance, and enhancing administrative efficiency. These advancements foster a more engaging and effective educational experience, preparing students better for the future.

The convergence of AI and IoT has also made strides in environmental conservation efforts. In Kenya, AIoT is employed for wildlife monitoring, using drones and IoT sensors to track animal movements and identify poaching activities. AI algorithms analyze the data to predict and prevent poaching incidents, playing a crucial role in wildlife conservation. These efforts have seen a positive impact on the preservation of endangered species and highlight the potential for technology to contribute actively to environmental sustainability.

In summary, the success stories stemming from real-world AIoT applications are numerous and span various domains. They illustrate how the synergy of AI and IoT can drive efficiency, enhance user experiences, and contribute to sustainable practices. As these technologies continue to evolve, they hold the promise of further transforming industries and improving the quality of life globally. These successes serve as a powerful reminder of what's possible and inspire continuous innovation in the realm of AIoT.

Lessons Learned

As we delve into the real-world applications of AIoT (Artificial Intelligence and Internet of Things), several key lessons emerge from the deployment and integration of these transformative technologies. These lessons are crucial for understanding both the potential and limitations of AIoT, making them invaluable for stakeholders across various industries. The convergence of AI and IoT has provided insights that help optimize operations, improve user experiences, and catalyze innovation. However, challenges remain that must be tackled to fully leverage the capabilities of AIoT.

First and foremost, one of the core lessons learned is the critical importance of data quality. The integration of AI and IoT relies heavily on data to deliver meaningful insights and drive intelligent actions. Poor data quality compromises the efficacy of AI algorithms, leading to erroneous conclusions or suboptimal decisions. Organizations have realized the necessity of establishing robust data governance frameworks to ensure the accuracy, consistency, and reliability of data. Clean and well-structured data is the bedrock upon which effective AIoT solutions are built.

Effective data integration also stands out as a vital point. AIoT systems pull data from various sources, often in real-time. Seamless data integration ensures that disparate data streams can be synthesized into coherent and actionable insights. Early adopters have grappled with the intricacies of integrating data from multiple sensors, devices, and platforms. This lesson underscores the need for standardized data protocols and interoperability to smoothen data amalgamation.

Security and privacy are paramount in the realm of AIoT. With the increasing number of connected devices and the sensitive nature of the data they collect, industries have learned that robust security measures are non-negotiable. Implementing advanced encryption techniques, continuous monitoring, and prompt vulnerability assessments are now standard practices. The lessons of high-profile security breaches have highlighted the need for a proactive rather than reactive approach to security. Privacy considerations, particularly in healthcare and finance, have taught organizations to be transparent about data usage and to ensure compliance with regulations such as GDPR and HIPAA.

Scalability is another lesson that has come to the fore. While pilot projects and initial deployments of AIoT solutions often yield promising results, scaling these solutions to enterprise or urban levels poses significant challenges. Successful scalability requires careful planning, robust infrastructure, and the flexibility to adapt to new requirements. Companies have learned to adopt an incremental approach, scaling up in manageable phases to identify and address roadblocks before they impede broader implementation.

The human factor cannot be overlooked. The integration of AI and IoT demands a skilled workforce capable of navigating complex technologies. Training and continuous education are vital to equip teams with the necessary skills. Moreover, cross-disciplinary collaboration often leads to the most innovative solutions. Organizations have learned to break down silos and foster collaboration between IT experts, data scientists, operations teams, and business strategists.

From a project management perspective, agility is an invaluable lesson. Traditional linear project management methodologies often fall short in the dynamic landscape of AIoT. Adopting agile practices allows teams to iterate quickly, respond to emerging trends, and pivot based on real-time feedback. This flexibility can significantly enhance the success rates of AIoT projects.

Another crucial lesson is the significance of user-centered design. AIoT technologies, no matter how advanced, must serve the end user's needs. Whether it's a consumer smart home solution or an industrial IoT application, the user interface should be intuitive and accessible. Feedback loops with end-users can provide invaluable insights that drive iterative improvements and enhance user satisfaction.

Resource management has also surfaced as a fundamental aspect of AIoT implementation. Efficient resource allocation ensures the optimized performance of AIoT systems while keeping operational costs in check. Insights from real-world applications suggest that predictive analytics can play a key role in resource management, from forecasting demand to preventing downtime through predictive maintenance.

Environmental sustainability is yet another lesson derived from AIoT applications. As industries increasingly integrate AIoT, they have become more aware of its potential to drive sustainable practices. Smart grids, AI in renewable energy, and efficient resource management are just a few examples where AIoT has enabled more sustainable operations. The lesson here is clear: AIoT can be a powerful ally in the fight against climate change and resource depletion.

One cannot overlook the societal impact of AIoT. Real-world applications have revealed the potential for AIoT to address public-good challenges such as urban congestion, healthcare access, and energy distribution. However, this potential must be balanced against ethical considerations. Societies have learned to examine who benefits from these technologies and to ensure equitable access and implementations that do not exacerbate existing inequalities.

Lest we forget, the resilience of AIoT systems in the face of disruptions is a crucial lesson. Unexpected events, such as natural disasters or cyber-attacks, can test the robustness of AIoT infrastructures. Lessons from past disruptions have encouraged industries to focus on building resilient systems capable of maintaining operations under adverse conditions and quickly recovering from any disruptions.

Finally, the ever-evolving regulatory landscape is a critical lesson for all stakeholders. As AIoT technologies advance, so do the regulatory frameworks governing them. Organizations have learned to stay agile and adaptable, ensuring compliance with existing regulations while anticipating future regulatory changes. This proactive stance mitigates the risks of legal challenges and fosters consumer trust.

In summary, the real-world applications of AIoT provide a treasure trove of lessons learned that span data quality, security, scalability, human factors, project management, user-centered design, resource management, sustainability, societal impact, resilience, and regulatory compliance. These lessons are not just academic; they offer practical insights that can guide the successful deployment and optimization of AIoT solutions across various sectors. As the journey of AIoT continues to unfold, staying observant and adaptable to these lessons will be crucial for harnessing its full potential.

CHAPTER 13:
ETHICAL AND PRIVACY
CONSIDERATIONS

As we dive deeper into the transformative world of AI and IoT, it's imperative to address the ethical and privacy concerns that come with it. The collection and analysis of vast amounts of data raise significant questions about consent, security, and the potential for misuse. Ensuring data security is not just a technical challenge but a moral imperative, safeguarding users from breaches that could lead to identity theft or unauthorized surveillance. Equally important is the commitment to ethical AI practices, where biases must be identified and mitigated to prevent discriminatory outcomes. Privacy concerns in IoT can no longer be an afterthought; they must be built into the design and deployment of these technologies. By prioritizing ethical considerations and privacy protections, we not only build trust with users but also pave the way for sustainable and equitable technological advancement. This holistic approach ensures that the benefits of AI and IoT can be fully realized without compromising the fundamental rights of individuals.

Data Security

As AI and IoT become increasingly intertwined, the importance of data security cannot be overstated. Given the vast amounts of data generated by IoT devices and the sophisticated analytics performed by AI, ensuring the security of this data is paramount. The risk of data breaches, unauthorized access, and data manipulation presents significant concerns for both individuals and organizations. These concerns stretch across various sectors from healthcare to finance, and the repercussions of a security lapse can be severe, ranging from financial losses to compromised personal safety.

Data security in the realm of AI and IoT involves multiple layers. At the foundational level, the data itself must be encrypted both in transit and at rest. This ensures that, even if intercepted, the data remains unintelligible without the proper decryption keys. Advanced encryption standards and secure communication protocols form the bedrock of protecting sensitive information as it moves through networks of IoT devices and AI systems.

Perimeter security is another critical aspect. Firewalls, intrusion detection systems, and gateways act as the first line of defense against unauthorized access. These tools monitor network traffic and look for any signs of unusual activity that might indicate a security threat. Even so, attackers are continually evolving their methods, necessitating robust and adaptive security strategies.

Beyond these external defenses, AI itself can be a powerful ally in securing data. AI systems can process vast amounts of security-related data, identifying patterns indicative of potential threats faster than any human could. By leveraging machine learning algorithms, these AI-based security systems can continually improve, learning from each attempted breach and becoming more effective over time. This makes for a dynamic and responsive security environment that can counteract threats in real-time.

However, with AI comes the responsibility of securing not only the data but also the AI models and algorithms themselves. Algorithmic integrity is a crucial facet of AI security. The models that interpretation and decision-making rely on must be shielded from tampering. Adversarial attacks, where malicious actors subtly alter the environment to mislead the AI's understanding, present a sophisticated threat. Ensuring algorithmic robustness involves rigorous testing, validation, and the implementation of safeguards against such tampering attempts.

The integration of IoT devices poses specialized challenges due to their diverse nature and decentralized architecture. Many IoT devices operate with limited computing power, making traditional security measures impractical. Consequently, lightweight encryption algorithms and low-footprint security protocols are vital. Additionally, ensuring that these devices can receive security updates is critical. Over-the-air (OTA) updates facilitate the regular implementation of patches, securing vulnerabilities that may be discovered post-deployment.

Identity and access management (IAM) is another cornerstone of data security within AI and IoT ecosystems. Assigning proper permissions and establishing strong authentication mechanisms are essential to ensuring that only authorized users and devices can access sensitive data. Multifactor authentication (MFA), biometric verification, and robust password policies aid in reinforcing security at the access control level. For IoT devices, securing the identity not just of human users but of the devices themselves is critical. Digital certificates and hardware-based secure elements can authenticate device identities, preventing unauthorized devices from joining networks.

Regulatory compliance also becomes a key consideration in the landscape of AI and IoT. Various jurisdictions have implemented stringent data protection laws, such as the General Data Protection Regulation (GDPR) in the European Union and the California Consumer Privacy Act (CCPA) in the United States. Compliance with these regulations demands comprehensive strategies that address data security and user privacy. This not only helps in avoiding legal ramifi-

Charlie Morgan

cations but also in building trust with users and customers who are increasingly aware of privacy issues.

Moreover, transparent data handling procedures enhance user trust. Transparency involves clear communication about what data is being collected, how it will be used, who will have access to it, and how long it will be retained. Users should have the ability to opt-in or out of data collection practices and control over their personal information. Ensuring transparent practices can also mitigate the risk of data misuse and strengthen the relationship between providers and consumers.

The collaborative nature of AI and IoT deployments also requires secure interfaces and APIs. These connections between systems should be fortified against unauthorized access and data interception. Implementing strong API security measures, such as rate limiting, authentication, and input validation, can prevent attacks that target these interfaces. Ensuring that all third-party services integrated with AI and IoT solutions adhere to stringent security standards is also important.

Network segmentation is another tactic that enhances data security. By segmenting a network into smaller parts, organizations can contain potential breaches to isolated sections, minimizing the overall impact. This is particularly beneficial in IoT environments where diverse devices operate within a single network. Each segment can be secured based on its specific requirements, and compromised devices can be quarantined without affecting the entire system.

Finally, the human element cannot be ignored. Training and awareness programs for employees and users are critical components of a comprehensive data security strategy. Human error, often in the form of weak passwords or phishing attacks, remains one of the most common causes of data breaches. Educating users on best practices in data security, recognizing common threats, and promoting a culture of vigilance significantly bolsters overall security posture.

In conclusion, data security within AI and IoT ecosystems is multi-faceted, requiring an approach that combines technical safeguards, regulatory compliance, user education, and continuous adaptation to evolving threats. As we continue to advance in the integration of AI and IoT, prioritizing the security of data will ensure that these technologies can achieve their transformative potential while safeguarding the trust and safety of individuals and societies at large.

Ethical AI Practices

In a world where AI and IoT technologies are becoming increasingly integrated into everyday life and industry, ethical AI practices are no longer just a consideration but a necessity. The deployment of AI in various IoT applications comes with profound responsibilities that go beyond technical execution. These responsibilities encompass a broad spectrum of ethical considerations, including fairness, accountability, transparency, and social impact. But what do these terms mean in practice, and how can businesses and developers ensure they're adhering to them?

First, let's delve into the concept of fairness. AI algorithms often make decisions that can significantly affect people's lives, from determining credit scores to diagnosing medical conditions. Ensuring that these algorithms are fair means they must be designed and tested to avoid biases that can result from skewed data sets or flawed logic. This, however, is easier said than done. Data used in training machine learning models often reflect existing societal biases, which can be magnified by the algorithm.

One way to address this is through diverse and inclusive data collection practices. By ensuring that data sets capture a wide range of experiences and conditions, developers can minimize the risk of biased outcomes. Another approach involves algorithmic auditing and bias detection tools that continually monitor and evaluate AI systems for fairness.

Accountability in AI is another crucial ethical practice. When AI systems make decisions, it should be clear who is responsible for those decisions. Is it the developers, the company deploying the technology, or the AI system itself? Establishing clear lines of accountability helps ensure that there are processes for addressing errors, making improvements, and providing recourse for individuals adversely affected by AI decisions.

Transparency goes hand in hand with accountability. AI systems should be designed in a way that their decision-making processes can be understood and scrutinized. This doesn't necessarily mean revealing proprietary algorithms but rather providing explanations for how decisions are made and what data influences them. This level of transparency and explainability can foster trust among users and stakeholders, ensuring that AI-driven decisions are perceived as legitimate and justifiable.

Let's not overlook the ethical implications of AI's social impact. AI and IoT technologies have the potential to create profound societal changes, both positive and negative. For instance, AI can help optimize resource use in smart cities, leading to more sustainable urban living. However, it could also lead to job displacement in sectors where automation becomes prevalent. Ethical AI practices must include strategies for mitigating these negative impacts, such as retraining programs and social safety nets for workers displaced by automation.

In the healthcare sector, the ethical application of AI can literally be a matter of life and death. AI systems that aid in diagnosis and treatment must be rigorously tested for accuracy and reliability. Moreover, they should complement rather than replace human judgment, ensuring that healthcare professionals retain the final say in medical decisions. This partnership between AI and human experts can lead to more informed and effective healthcare, benefiting patients and providers alike.

While AI offers remarkable potential for advancements, it also raises questions about privacy and informed consent. Users should be fully aware of how their data is being collected, used, and shared. Transparent privacy policies and consent mechanisms are crucial for ensuring that individuals maintain control over their personal information. This includes providing users with clear options to opt-out of data collection and understanding the long-term implications of data sharing.

Moreover, ethical AI practices require a collaborative approach involving multiple stakeholders. Governments, private sector companies, academic institutions, and civil society organizations all play a role in shaping the ethical framework for AI and IoT. Policymakers must develop regulations that protect public interests without stifling innovation. Businesses should commit to ethical standards and practices that prioritize societal good along with profitability. Academia provides the critical research and thought leadership necessary for understanding the complex ethical issues related to AI.

Educational initiatives are also vital in fostering an environment where ethical AI practices thrive. There's a need to equip current and future AI professionals with the skills and knowledge to recognize and address ethical issues. This means integrating ethics into AI and computer science curricula and encouraging interdisciplinary studies that include social sciences and humanities perspectives. By doing so, we can develop AI practitioners who are not only technically proficient but also ethically aware and socially responsible.

An essential component of ethical AI practices is the principle of inclusivity. AI systems should be designed to benefit a broad spectrum of the population, avoiding the creation or reinforcement of inequalities. This requires inclusive design practices and engaging with diverse communities to understand their unique needs and challenges. By prioritizing inclusivity, AI and IoT can become powerful tools for social good, promoting equity and justice.

Lastly, ethical AI practices must be understood as an ongoing commitment rather than a one-time checklist. The field of AI is rapidly evolving, and new ethical questions will undoubtedly arise. Continuous dialogue, evaluation, and adaptation are necessary to keep pace with technological advancements and societal changes. Organizations should establish ethical review boards and conduct regular assessments of their AI systems to ensure they remain aligned with ethical standards.

In sum, ethical AI practices are multifaceted and require a holistic approach, blending technical, social, and regulatory considerations. By committing to fairness, accountability, transparency, and social responsibility, we can harness the full potential of AI and IoT to create a better, more equitable world. The integration of AI in IoT offers unprecedented opportunities, but it also demands a conscientious and ethical approach to ensure that these technologies serve humanity's best interests.

Privacy Concerns in IoT

The Internet of Things (IoT) is transforming the way we interact with the world, connecting everyday objects to the internet and to each other. While this integration offers unprecedented convenience and efficiency, it also raises significant privacy concerns that must be addressed. The sheer volume of data generated by IoT devices is staggering. Every smart thermostat, wearable fitness tracker, and connected car consistently collects and transmits data, creating a rich tapestry of information about our daily lives. But who is watching over this data? And how secure is it?

Primarily, the data collected by IoT devices often includes sensitive personal information. This could range from health metrics gathered by wearable health devices to location data tracked by smart home systems. The central issue here is informed consent. Often, users are not fully aware of the extent of the data being collected, nor do they understand how this data will be used or shared. When downloading a new app or setting up a new device, it's common to encounter lengthy privacy policies written in complex legal language. Few take the time to read through these documents, let alone fully comprehend them. This lack of transparency can lead to unintentional privacy breaches.

Another significant concern is data vulnerability. IoT devices can be an attractive target for hackers, given their broad adoption and the volume of data they handle. Many devices are designed with convenience rather than security in mind, leaving gaps that can be easily exploited. Could a hacker take control of your smart thermostat, or worse, gain access to your home's security system? Unfortunately, the answer is yes. The implications of such breaches can be profound, impacting personal safety and financial security.

Compounding the issue, many IoT devices operate on outdated software. Due to cost or lack of awareness, manufacturers may not provide regular updates or patch vulnerabilities in their products. As a result, devices that are otherwise secure at the time of purchase can quickly become compromised as new threats emerge. The responsibility often falls on consumers to ensure their devices are regularly updated, but this is a task many are neither equipped nor inclined to manage.

Moreover, the aggregation of data from various IoT devices can result in invasive profiling. Imagine an ecosystem where your daily activities, shopping habits, health information, and even social interactions are collected, analyzed, and potentially sold to third-party advertisers. Such comprehensive profiling can lead to hyper-targeted advertising, which some find intrusive or manipulative. Beyond commercial use, there's a risk that aggregated data could be accessed by governments or other entities, potentially infringing on individual freedoms and privacy.

The concept of data ownership also comes into play. With IoT devices, it's often unclear who owns the data being generated. Is it the manufacturer who created the device, the service provider who processes the data, or the user who generates it? This ambiguity can lead to disputes and a lack of accountability when data is misused or mishandled. Establishing clear guidelines and ownership rights is paramount to addressing these concerns.

In addition to individual privacy, IoT also impacts privacy on a broader scale. For instance, in smart cities where IoT devices are used for traffic management, public safety, and resource distribution, the data collected can paint a detailed picture of urban life. While this information can optimize city function and improve residents' quality of life, it also exposes the population to mass surveillance. How do we balance the benefits of a connected city with the right to privacy for its residents?

The principle of data minimization can be a potent countermeasure against these privacy concerns. By ensuring that only necessary data is collected and stored, and for the minimum duration required, the potential damage from data breaches can be mitigated. Similar to this, anonymizing data to the greatest extent possible can add an additional layer of protection, ensuring that even if data is intercepted or misused, it cannot be easily traced back to individuals.

Regulations and laws play a critical role in protecting privacy within the IoT ecosystem. The General Data Protection Regulation (GDPR) in the European Union is one such legal framework that strives to give individuals control over their personal data while mandating stricter data protection measures for companies. However, global consistency in such regulations is lacking. There's a need for more cohesive and comprehensive policies worldwide that address the unique challenges posed by IoT.

Ethical considerations also arise with the development and deployment of IoT technologies. It's essential for companies to adopt ethical practices in data collection, usage, and sharing. This includes obtaining explicit consent, being transparent about data practices, and providing users with easy-to-understand privacy options. Ethical AI practices, as discussed elsewhere in this book, are integral to handling IoT data responsibly and should go hand-in-hand with technical solutions to secure data.

As we navigate these privacy challenges, it's important to keep the user at the center of every discussion. Prioritizing user education can empower individuals to make informed decisions about their privacy. This encompasses clear communication about data practices, straightforward privacy settings, and easy access to delete personal data. By making privacy a key feature rather than an afterthought, we can foster greater trust and acceptance of IoT technologies.

Technological solutions such as blockchain also offer promising avenues for enhancing privacy in IoT. Blockchain's decentralized nature can provide transparency in data transactions and robust security measures. By ensuring that data is encrypted and only accessible to authorized entities, blockchain can implement a layer of trust in the IoT ecosystem. Nonetheless, the effectiveness and scalability of such solutions remain to be fully explored.

Lastly, industry collaboration is crucial. Manufacturers, service providers, lawmakers, and privacy advocates must work together to develop standards and best practices for IoT privacy. This collaboration can lead to the creation of universally accepted protocols that ensure robust privacy protections across the board. Through forums, working groups, and partnerships, the industry can collectively address the privacy concerns that accompany the rapid growth of IoT.

Ensuring privacy in the ever-expanding landscape of IoT is undeniably complex but achievable. Through the combination of regulatory measures, technological innovations, ethical practices, and user empowerment, the coherence of privacy in an interconnected world can be strengthened. As IoT continues to evolve, vigilance and proactive measures will be key to safeguarding personal privacy and fostering an environment of trust and innovation.

CHAPTER 14:
CHALLENGES AND LIMITATIONS

Despite the transformative potential of integrating AI and IoT, several challenges and limitations remain. Technical hurdles, like ensuring robust data collection and handling, can impede the seamless functionality of AIoT systems. Interoperability issues arise when diverse devices and platforms struggle to communicate effectively, often due to differing standards and protocols. Additionally, scalability presents significant obstacles, as expanding AIoT applications to a broader scope demands substantial investments in infrastructure and technology. Addressing these challenges collectively is crucial for harnessing the full benefits of AIoT, requiring collaborative efforts from industry stakeholders, researchers, and policymakers.

Technical Hurdles

As we journey through the integration of AI and IoT, it's impossible to ignore the technical hurdles that come with merging these two transformative technologies. Despite the monumental potential, the road is laden with significant challenges that require innovative thinking and advanced engineering solutions. These hurdles aren't merely speed bumps; they are intricate roadblocks that need to be addressed for successful implementation and widespread adoption.

Among the primary technical hurdles is the sheer complexity inherent in integrating AI and IoT devices. Both technologies involve a myriad of components, from sensors and networks to machine learning algorithms and analytical models. Getting these components to work seamlessly can be a daunting task. The problem becomes more pronounced when dealing with legacy systems that were never designed to be compatible with modern AI algorithms. Engineers often find themselves wrestling with outdated hardware and software that lack the computational power and flexibility to handle AI workloads efficiently.

Additionally, issues related to data collection, storage, and analysis present substantial obstacles. IoT devices generate vast amounts of data, which need to be stored securely and processed in real-time. Handling this data deluge requires robust and scalable cloud infrastructure, capable of supporting high-speed data transfer and storage. This necessitates significant investment in both on-premises and cloud-based solutions, which can be a barrier for smaller organizations. Furthermore, ensuring the integrity and consistency of this data as it moves across various platforms and devices adds another layer of complexity.

Power consumption is another technical hurdle that cannot be overlooked. IoT devices, especially those deployed in remote or hard-to-reach locations, are often battery-powered and require energy-efficient designs to function effectively over extended periods. Integrating AI capabilities into these devices further complicates the matter, as machine learning algorithms typically require significant computational resources. Achieving a balance between performance and power efficiency is crucial for the sustainability of AI-powered IoT solutions.

Communication protocols also present significant challenges. The myriad of IoT devices use different communication standards and protocols, leading to interoperability issues. For example, a sensor might use Bluetooth, while another device might rely on Zigbee or Wi-Fi. Ensuring that all these devices can communicate effectively in a heterogeneous environment requires advanced networking solutions and standards. Furthermore, the development of universal protocols that can accommodate the diverse requirements of various IoT devices is still in its nascent stages, adding another hurdle to seamless integration.

Latency and real-time processing represent another set of hurdles. Many AIoT applications, such as autonomous vehicles and industrial automation, require real-time data processing and decision-making. However, latency in data transmission and processing can severely impact the performance and reliability of these applications. Factors such as network congestion, physical distance between devices and data centers, and the computational load can contribute to unacceptable delays. Addressing these issues requires the deployment of edge computing solutions, where data processing occurs closer to the data source, thus reducing latency.

Security concerns are paramount when dealing with AI and IoT integration. The massive volumes of data generated by IoT devices are a goldmine for cybercriminals. Protecting this data from breaches and unauthorized access is crucial, yet challenging. Traditional security measures often fall short when applied to the AIoT ecosystem, necessitating the development of advanced security protocols and encryption techniques. Moreover, the distributed nature of IoT networks makes them susceptible to a range of attacks, including DDoS (Distributed Denial of Service) attacks, which can cripple entire systems.

Another technical hurdle is the need for high reliability and uptime. Many AIoT applications are mission-critical, where failure is not an option. This is particularly true in sectors like healthcare, transportation, and industrial automation. Ensuring the reliability of these systems involves rigorous testing, real-time monitoring, and robust fault-tolerant designs. Additionally, AI models themselves need to be highly accurate and reliable, as errors in prediction or decision-making can have dire consequences. This often requires continuous learning and model updating, which adds to the complexity.

Training AI models to work efficiently with IoT data is far from straightforward. IoT data is often noisy, incomplete, or imbalanced, which poses significant challenges for training accurate and robust AI models. Data preprocessing techniques such as cleaning, normalization, and augmentation are essential but can be resource-intensive and time-consuming. Moreover, the computational requirements for training these models are substantial, necessitating high-performance computing resources.

Another significant hurdle is the technical debt that can accrue when rapidly developing and deploying AIoT solutions. Companies often rush to bring products to market to stay ahead of the competition, which can lead to shortcuts and compromises in design and implementation. Over time, this technical debt can become a significant burden, complicating future updates and scaling efforts. Addressing this issue requires a careful balance between speed and quality, with an emphasis on robust software engineering practices from the outset.

Finally, there is the human element to consider. Developing and maintaining AIoT systems requires a highly skilled workforce with expertise in diverse fields such as machine learning, data science, electrical engineering, and software development. Finding individuals with this combination of skills can be challenging, and the demand for such talent is high. Moreover, continuous training and development are necessary to keep pace with rapidly evolving technologies. Addressing this talent gap is crucial for overcoming the technical hurdles in AIoT integration.

In sum, the technical hurdles in integrating AI and IoT are multifaceted and complex. They span across hardware and software, data management, power efficiency, communication protocols, latency and real-time processing, security, reliability, AI model training, technical debt, and the human element. Overcoming these challenges requires a concerted effort from engineers, researchers, and organizations, along with significant investment in new technologies and infrastructure. However, the rewards are well worth the effort, as successful integration holds the promise to revolutionize industries, enhance everyday life, and shape the future of technology in ways we can only begin to imagine.

Interoperability Issues

When integrating AI and IoT, one of the most challenging obstacles is interoperability. Interoperability refers to the ability of different systems, devices, or applications to communicate and work together seamlessly. In a rapidly advancing technological ecosystem, achieving this harmony can be incredibly complex. The vast array of devices, platforms, and communication protocols can lead to significant hurdles that need to be addressed for the successful implementation of AIoT solutions.

Various devices in the IoT ecosystem come from different manufacturers, each with its own proprietary technology and standards. This disparity leads to compatibility issues, making inter-device communication problematic. For instance, consider a smart home environment where devices like thermostats, security cameras, and light bulbs need to interact. If these devices operate on different protocols, harmonizing them into a single, cohesive system becomes an arduous task.

Another layer of complexity arises from the diverse range of communication protocols employed in IoT systems. Protocols such as Zigbee, Z-Wave, Bluetooth, and Wi-Fi each have distinct advantages and limitations. Yet, they often don't naturally interact with one another, necessitating additional hardware or software solutions to bridge the gaps. These bridging solutions can introduce latency and potential points of failure, further complicating an already intricate system.

Software interoperability is equally challenging. AI algorithms and IoT platforms often operate in silos, utilizing different programming languages, data formats, and machine learning models. For example, an AI algorithm designed to analyze smart grid data may be written in Python using specific libraries, while the IoT platform collecting the data might use a different technology stack entirely. Translating and harmonizing these diverse elements into a unified system often requires significant customization and sophisticated middleware solutions.

Data interoperability is another critical concern. IoT devices generate massive amounts of data, but this data is valuable only if it can be effectively shared and analyzed. Data formats and standards differ, causing difficulties in data integration. Structured data from sensors may not easily integrate with unstructured data from voice assistants, for instance. Establishing universal data standards and formats could simplify this process, but achieving consensus and widespread adoption is an ongoing challenge.

Moreover, security and privacy issues compound interoperability challenges. Each connection point between devices can potentially serve as a security vulnerability, requiring robust encryption and secure communication protocols to safeguard data integrity. As systems become more interconnected, ensuring that all components meet stringent security standards is vital to protect against cyber threats. This necessitates not only technological solutions but also industry-wide protocols and regulations.

On the organizational front, companies often have to balance the cost and complexity of achieving interoperability with the potential benefits. Retrofitting legacy systems to be compatible with new AIoT solutions can be prohibitively expensive. Organizations must weigh whether the investment in interoperability will yield a significant return on investment in terms of efficiency, productivity, and innovation.

Vendor lock-in exacerbates these issues further. When organizations commit to a specific vendor's ecosystem, they may find it challenging to integrate devices or systems from other vendors. This lock-in limits flexibility and can stifle innovation, as organizations are constrained to the features and updates provided by a single vendor. This problem underscores the need for open standards and frameworks that promote cross-vendor interoperability, encouraging a more dynamic and flexible technological landscape.

There are initiatives underway aimed at tackling these interoperability issues. Industry consortia and standard-setting bodies are working to develop unified communication protocols and interoperability frameworks. For example, the Open Connectivity Foundation (OCF) and the Industrial Internet Consortium (IIC) are making strides in establishing standards that promote cross-device communication and integration. These efforts, while promising, require widespread adoption and active participation from various stakeholders to be truly effective.

Interoperability isn't just a technical challenge; it's also a business and strategic issue. Organizations need to establish clear strategies and frameworks for dealing with these challenges. This involves not only selecting the right technology stack but also fostering partnerships and ecosystems that facilitate seamless integration. Collaboration between technology providers, standard-setting bodies, and end-users is essential to create an environment where interoperability is not seen as an afterthought but as a foundational aspect of AIoT deployment.

Intriguingly, the concept of digital twins is emerging as a potential solution to some of these interoperability issues. Digital twins are virtual replicas of physical devices that can simulate and predict their real-world counterparts' behavior. By using digital twins, organizations can create a unified interface for disparate systems, simplifying data integration and analysis. However, while promising, the implementation of digital twins also brings its own set of challenges, particularly around data accuracy and computational requirements.

The future of AIoT hinges on overcoming these interoperability hurdles. As technology continues to advance and the landscape of devices and platforms expands, the importance of seamless integration will only grow. Achieving this will require a concerted effort from all sectors involved, including technology vendors, regulatory bodies, and end-users. By addressing interoperability head-on, the potential for AIoT to revolutionize industries, enhance everyday life, and shape the future of technology becomes increasingly attainable.

In summary, while the path to achieving interoperability in AIoT is fraught with challenges, it also presents immense opportunities. Through collaborative efforts, innovative solutions, and a commitment to open standards, we can pave the way for a more interconnected and efficient technological ecosystem. The potential benefits are profound, promising a future where devices and systems work together seamlessly to deliver enhanced functionality, improved efficiency, and transformative advancements in numerous fields.

Scalability Challenges

The vision of integrating AI with IoT into a seamless and ubiquitous network offers enormous potential. But, with great potential comes equally significant hurdles, particularly in the realm of scalability. As industries and daily life increasingly depend on integrated AIoT solutions, the necessity to scale these systems to manage growing demands cannot be overstated. Yet, the journey to achieve seamless scalability is fraught with complex challenges.

At the core, one of the primary scalability challenges is the sheer volume of data generated by IoT devices. As IoT networks expand, the data collected escalates exponentially. This overwhelming influx can put immense pressure on storage, processing, and analysis capabilities. Processing this vast amount of data in real-time is not a mere technical feat; it demands sophisticated algorithms, advanced computing infrastructure, and considerable bandwidth.

Moreover, data heterogeneity poses a significant scalability issue. IoT devices produce diverse data types, from sensor readings and video feeds to audio signals and environmental metrics. Standardizing and integrating these disparate data forms into a cohesive system requires robust data management and harmonization strategies. Without effective methods to unify and process varied data streams, scaling becomes an elusive goal.

Network connectivity is another critical aspect. As more IoT devices are deployed, the network's ability to support increased traffic becomes a bottleneck. Robust, high-speed, and reliable network infrastructure is paramount to accommodate the growing number of connected devices. Ensuring seamless connectivity and reducing latency is essential for the effective functioning of AI applications that rely on real-time data processing and decision-making.

Scalability isn't just about handling more devices or more data; it's also about maintaining system integrity. As AIoT systems grow, so does the complexity of managing these systems. Ensuring system reliability, minimizing downtime, and efficiently managing resources become challenging tasks. Scalability efforts must encompass robust system monitoring and management tools to preemptively address potential failures and optimize performance.

Furthermore, security concerns magnify with scale. More devices mean more potential entry points for cyber threats. As the AIoT ecosystem expands, ensuring robust security protocols to safeguard data and systems is imperative. This includes implementing encryption, secure communication channels, and proactive threat detection mechanisms. As devices proliferate, consistently updating and managing security across a wide array of devices becomes a daunting task.

Another layer of complexity is the integration of legacy systems with new IoT and AI technologies. Many industries have existing infrastructures that might not be inherently compatible with modern AIoT solutions. Retrofitting these systems to enable smooth interoperability without compromising performance or security requires significant effort and investment. This integration challenge involves both technical adaptations and training personnel to manage these hybrid environments effectively.

The rapid pace of technological evolution adds to the pressure. Emerging technologies and standards continually shift the landscape. Keeping AIoT solutions up-to-date with the latest advancements while scaling operations can be a heavy lift. It necessitates flexible architectures and adaptive systems that can evolve without extensive overhauls, ensuring long-term scalability and relevance.

In addition to technical challenges, organizational readiness also plays a crucial role. Adequately skilled personnel are required to design, implement, and manage scalable AIoT systems. This means investing in training and development to build a knowledgeable workforce capable of tackling the complexities of scalability. Moreover, organizational culture needs to be receptive to change, encouraging innovation and continual improvement.

Another economic aspect is cost management. Building scalable AIoT solutions often entails significant upfront investments in infrastructure, software, and talent. Striking a balance between investing in scalability and managing costs is a delicate act. Organizations must carefully plan their scaling strategies to achieve cost-effective scalability without compromising on performance or quality.

To compound the scalability challenge, regulatory compliance becomes more intricate as AIoT systems scale. Different regions and industries have varied regulatory requirements. Ensuring compliance across a widespread network of devices and data streams necessitates diligent monitoring and implementation of regulatory practices. Non-compliance can not only lead to legal penalties but also erode trust and credibility.

Ultimately, overcoming these scalability challenges demands a multidimensional approach. Innovations in edge computing, for example, can alleviate some of the burdens by processing data closer to the source, reducing latency, and lowering data transfer volumes. Additionally, leveraging cloud-based platforms can offer flexible and scalable storage and processing capabilities, supporting the vast data needs of AIoT systems.

Collaboration and shared knowledge can also be formidable tools in addressing scalability. Partnerships between technology providers, industry stakeholders, and academic institutions can drive the development of innovative, scalable AIoT solutions. Open-source platforms and community-driven initiatives can accelerate progress by pooling collective expertise and resources.

As AI and IoT continue to converge and permeate various facets of life and industry, scalability challenges will remain a dynamic landscape requiring continuous adaptation and innovation. The journey towards scalable AIoT solutions is not a finite destination but an ongoing evolution, where each advance paves the way for the next leap forward. With diligent efforts and collaborative spirit, the scalability hurdles can be transformed into stepping stones, leading to a future where the full potential of AIoT is realized across an ever-expanding horizon.

CHAPTER 15:
THE FUTURE OF AIOT

The future of AIoT is incredibly promising, characterized by rapid technological advancements and innovative applications that will transform how we live and work. Emerging trends are steering us towards more interconnected and intelligent systems, where AI and IoT not only co-exist but enhance each other synergistically. Imagine a world where smart cities optimize energy usage in real-time, healthcare systems predict and prevent diseases, and autonomous vehicles communicate seamlessly with intelligent infrastructure. These are no longer distant fantasies but imminent realities. Technological innovations like edge computing and 5G are poised to accelerate these developments, making AIoT more efficient and accessible. As we look forward, the predictions for AIoT are nothing short of revolutionary—anticipating a fully integrated ecosystem that personalizes experiences, augments human capabilities, and streamlines complex industrial processes. This exciting frontier holds the potential to create unprecedented opportunities across various sectors, driving both economic growth and societal advancement.

Emerging Trends

As we look towards the future, it's clear that the convergence of artificial intelligence (AI) and the Internet of Things (IoT) will fundamentally transform technology. Several emerging trends are making waves, signaling groundbreaking advancements in various industries. Understanding these trends is crucial to grasp the full potential of AIoT.

One significant trend is the rise of edge computing. With the proliferation of IoT devices, processing data at the network's edge—close to where it is generated—has become increasingly important. Edge computing minimizes latency, reduces the reliance on centralized data centers, and enhances real-time decision-making. This trend aligns perfectly with the capabilities of AI, as machine learning algorithms can be deployed on edge devices to analyze data instantly and take immediate action.

Another noteworthy trend is the evolution of 5G technology. 5G networks promise to boost internet speeds and greatly reduce latency, which are critical components for the effective deployment of AIoT solutions. 5G will enable more reliable and efficient communication between IoT devices, allowing for smoother and faster data transfer. This evolution supports more robust AI applications in areas such as autonomous vehicles, remote healthcare, and smart cities, where real-time data is essential.

AI-driven automation is also seeing a surge, finding its way into broader applications beyond traditional industrial uses. In retail, AIoT systems are automating inventory management and customer service, enhancing the shopping experience. In agriculture, AI-enabled drones and sensors are optimizing crop yields, monitoring soil health, and predicting pest infestations. This level of automation is driving efficiency and productivity across sectors.

Additionally, predictive analytics is becoming a staple in AIoT ecosystems. Leveraging historical data, AI models can predict future trends and behaviors with a high degree of accuracy. This capability is transforming fields such as predictive maintenance in manufacturing, where it can foresee equipment failures and schedule preemptive maintenance. Likewise, in healthcare, predictive analytics can identify potential health issues before they become critical, enabling early intervention and more personalized care.

Another trend to watch is the growing importance of cybersecurity in AIoT frameworks. As the number of connected devices explodes, so too does the vulnerability to cyber-attacks. Developing robust security protocols, leveraging AI for threat detection, and establishing secure communication channels are becoming paramount. Companies are increasingly focusing on integrating AI-driven security measures to safeguard their IoT networks from potential threats.

Ethical AI and responsible data usage are also gaining prominence. As AIoT solutions become more integrated into daily life, the ethical implications of data use, bias in AI models, and user consent are under scrutiny. Businesses and policymakers are being urged to adopt ethical guidelines to ensure transparency, fairness, and respect for user privacy. This shift towards ethical AI is not just a moral imperative but also a business necessity to build public trust and ensure long-term sustainability.

Furthermore, AIoT is facilitating more sustainable practices. From smart grids optimizing energy usage to IoT sensors monitoring environmental conditions, technology is playing a crucial role in driving sustainability. AI algorithms are helping industries reduce their carbon footprint, manage resources more efficiently, and adopt more sustainable business practices. This trend not only addresses ecological concerns but also meets the increasing demand from consumers and regulators for greener operations.

Voice-activated assistants and natural language processing (NLP) technologies are becoming increasingly sophisticated. AI-driven voice assistants are making IoT devices more accessible and user-friendly, enabling seamless interactions between humans and machines. These technologies are being adopted in smart homes, cars, and even industries, revolutionizing how users interact with their environment.

Blockchain technology is another emerging trend in the AIoT ecosystem. Blockchain offers a secure, decentralized method for recording transactions and managing data. By combining blockchain with AIoT, businesses can enhance data security, ensure transparency, and streamline processes. This union holds promise for various applications, including supply chain management, healthcare data security, and financial transactions.

There's also an interesting trend of digital twins in AIoT applications. Digital twins are virtual replicas of physical systems that can be analyzed and optimized using AI. In industries like manufacturing, digital twins aid in simulating processes, predicting equipment behavior, and improving operational efficiency. This concept is expanding to other sectors such as urban planning, healthcare, and even personal health management.

Moreover, context-aware systems are a growing trend in AIoT. These systems use real-time data from IoT sensors to understand an environment and make intelligent decisions. For example, in smart cities, context-aware traffic lights can adjust signal timings based on real-time traffic conditions, improving flow and reducing congestion. Such systems exemplify the potential of AIoT to enhance situational awareness and respond dynamically to changing conditions.

Hybrid AI models that combine multiple types of AI, such as traditional machine learning, deep learning, and reinforcement learning, are being developed to tackle more complex problems. These hybrid models are particularly useful in scenarios requiring flexibility and adaptability, such as autonomous driving, personalized healthcare, and adaptive learning environments.

Personalization is another area where AIoT is making significant strides. By collecting and analyzing data from various IoT devices, AI systems can tailor experiences to individual preferences and behaviors. From personalized shopping recommendations to bespoke healthcare plans, this trend highlights the potential of AIoT to improve user satisfaction and deliver customized solutions.

Finally, the integration of AIoT with augmented reality (AR) and virtual reality (VR) is an exciting frontier. These immersive technologies, powered by AI and fed with real-time IoT data, are finding applications in training, remote assistance, and even entertainment. For instance, in manufacturing, AR can overlay virtual instructions onto physical equipment for maintenance, while in retail, VR can offer immersive shopping experiences.

In conclusion, the emerging trends in AIoT are creating a dynamic landscape filled with opportunities and challenges. From edge computing to ethical considerations, each trend plays a crucial role in shaping the future of technology. As AI and IoT continue to evolve, they will undoubtedly unlock new possibilities, driving innovation and transforming industries. Staying abreast of these trends is essential for anyone looking to harness the full potential of AIoT.

As we move forward, it is crucial to remain adaptable and forward-thinking, ready to leverage the advancements in AIoT for a better, smarter, and more connected world.

Technological Innovations

The convergence of Artificial Intelligence (AI) and the Internet of Things (IoT), often termed AIoT, is setting the stage for a technological revolution that promises to reshape industries, redefine daily life, and drive leaps in efficiency and capability. The integration of these two transformative technologies is fueling a wave of innovation, where devices not only communicate but learn, adapt, and make autonomous decisions. This fusion augments the potential of IoT by infusing it with the cognitive power of AI, creating smarter, more responsive, and predictive systems.

At the heart of these technological innovations is edge computing. Edge computing brings computational power closer to where data is generated, enhancing real-time data processing and reducing latency issues traditionally associated with cloud computing. This is particularly critical for applications requiring instantaneous decision-making, such as autonomous vehicles and industrial automation. Through edge AI, data is processed locally on the devices themselves or nearby, negating the delay of sending data to centralized servers for processing. This advancement not only ensures faster response times but also enhances the privacy and security of data, addressing one of the fundamental challenges in IoT environments.

Machine learning models, integral to AIoT, are continuously evolving. One remarkable area of development is federated learning, a distributed approach that trains an algorithm across multiple decentralized devices holding local data samples, without exchanging them. This method enhances privacy and security by ensuring that the data remains on local devices. In the context of IoT, federated learning allows the vast network of devices to learn collectively while maintaining data confidentiality, which is paramount in industries like healthcare and finance where sensitive data is prevalent.

Another groundbreaking innovation is the advancement of neural networks, particularly deep learning, which enables systems to learn from vast amounts of data and make more accurate predictions and decisions. Innovations in neural network architectures, such as transformers and generative adversarial networks (GANs), are now being applied to IoT data to derive insights that were previously unattainable. For example, in the realm of smart cities, deep learning algorithms can analyze patterns from a multitude of connected sensors to optimize urban planning, manage traffic congestion, and even predict maintenance needs for municipal infrastructure.

Natural language processing (NLP), a subset of AI focusing on the interaction between computers and humans through natural language, is pivotal in enhancing user interaction with IoT devices. Voice-activated assistants like Amazon's Alexa and Google Assistant are prime examples of how NLP is making IoT devices more intuitive and user-friendly. These systems are becoming increasingly sophisticated, allowing for more natural and seamless interaction, which is critical for user adoption and satisfaction. Moreover, advancements in sentiment analysis and contextual understanding are ushering in a new era of personalized user experiences that adapt to individual preferences and needs.

In the industrial sector, advancements in robotics, powered by AIoT, are streamlining operations and maximizing efficiency. Collaborative robots, or cobots, are designed to work alongside human workers, enhancing productivity without compromising safety. These cobots are equipped with sensors and AI capabilities that enable them to learn from human actions, adapt to changing environments, and perform complex tasks with precision. In manufacturing, AI-driven predictive maintenance is reducing downtime and operational costs by analyzing data from connected machinery to predict failures before they occur, ensuring timely maintenance interventions.

In agriculture, AIoT is driving significant technological innovations through precision farming practices. By leveraging data from field sensors, drones, and satellite imagery, AI algorithms can analyze factors such as soil health, crop growth, and weather conditions to optimize farming practices. These insights enable farmers to make data-driven decisions about irrigation, fertilization, and pest control, resulting in increased crop yields and sustainable farming practices. Moreover, IoT-enabled equipment, like autonomous tractors and harvesters, further augments efficiency and productivity in agricultural operations.

The healthcare industry is witnessing transformative advancements with the integration of AIoT. Wearable devices, embedded with IoT sensors, continuously monitor patients' vitals and can detect health anomalies in real-time. Coupled with AI algorithms, these devices can predict potential health issues and alert healthcare professionals for timely intervention. Innovations in medical imaging, powered by AI, are also enhancing diagnostic accuracy, while telemedicine platforms are making healthcare more accessible by remotely connecting patients with doctors. These advancements are revolutionizing patient care and management, moving towards a more proactive and personalized healthcare system.

In the energy sector, smart grids and AI-driven energy management systems are revolutionizing how energy is distributed and consumed. Smart grids leverage IoT devices to monitor and manage energy flow, while AI algorithms optimize the balance between energy supply and demand. This not only enhances the reliability and efficiency of power systems but also accommodates the integration of renewable energy sources. Predictive analytics in energy consumption patterns are enabling more efficient resource management, reducing wastage, and promoting sustainability. Furthermore, AIoT solutions are being employed in the monitoring and maintenance of energy infrastructure, ensuring operational efficiency and minimizing downtime.

The retail industry is also experiencing a wave of technological innovations driven by AIoT. Smart shelves equipped with RFID tags and IoT sensors automate inventory management, providing real-time updates on stock levels and reducing the likelihood of out-of-stock scenarios. AI-driven analytics are enhancing customer experience by personalizing shopping recommendations based on purchasing behavior and preferences. Additionally, automated checkout systems, powered by computer vision, are transforming the shopping experience by allowing customers to make purchases without traditional checkout lines, vastly improving convenience and reducing waiting times.

Transportation and logistics are being transformed by autonomous vehicles and AI-powered fleet management systems. Self-driving cars and trucks, equipped with a multitude of sensors and AI algorithms, are revolutionizing the way goods and people are transported. These vehicles can navigate complex environments, avoid obstacles, and make real-time decisions, improving safety and efficiency. In logistics, AIoT is optimizing routes, managing fleets, and monitoring cargo conditions, ensuring timely and efficient delivery of goods. The integration of these technologies is paving the way for a more connected and efficient transportation infrastructure.

In finance and banking, AIoT is enhancing fraud detection and customer service. IoT-enabled payment systems, coupled with AI algorithms, monitor transactions in real-time, detecting and preventing fraudulent activities. Chatbots and virtual assistants, powered by AI, are improving customer service by providing instant support and resolving queries efficiently. Moreover, AI-driven analytics are offering personalized financial advice and insights, helping customers make informed decisions about their finances. These innovations are not only enhancing security but also improving the overall banking experience.

Beyond these specific applications, the ongoing advancements in AIoT are fostering an ecosystem of interconnected devices and systems that can autonomously optimize operations, drive efficiencies, and create new opportunities across various sectors. Innovations in AI algorithms, sensor technologies, and connectivity solutions are continually evolving, pushing the boundaries of what is possible. As these technologies mature, they are expected to become more accessible, affordable, and integrated into our daily lives, heralding an era of unprecedented technological transformation.

Moreover, the rise of open-source AIoT platforms is democratizing innovation, enabling developers and businesses to experiment, innovate, and bring new solutions to market faster than ever before. These platforms provide standardized tools and frameworks that simplify the development and deployment of AIoT applications, fostering

Future Predictions

The journey of AIoT is only just beginning, and the future holds immense possibilities that promise to revolutionize our lives even further. As technology continues to advance at an exponential rate, the integration of AI and IoT will become more seamless, embedding intelligence into every imaginable object and area in our daily lives and industries. The fusion of AIoT is set to redefine not just how we live and work, but the very fabric of our societies.

One of the most exciting prospects for the future of AIoT is in the realm of smart environments. Smart cities will evolve into intelligent ecosystems where every aspect is optimized for efficiency and sustainability. Urban planning will benefit from predictive analytics to manage resources better, anticipate traffic patterns, and improve public services. The notion of a "smart city" will extend to rural areas, ensuring that the benefits of AIoT are not limited to urban locales. The infrastructure will adapt in real-time, making cities not just connected but truly intelligent.

In the healthcare sector, the advances in AIoT will drive a paradigm shift towards predictive and preventive medicine. Personalized healthcare will become standard practice, powered by AI algorithms that can predict health issues before they occur and IoT devices that monitor patient vitals continuously. The future may even see the development of nano-IoT devices that operate at a cellular level, providing unprecedented insights into human health. This shift will not only improve the quality of care but also significantly reduce healthcare costs.

The industrial sector will witness an overhaul with advancements in AIoT, pushing the boundaries of what is possible in manufacturing, supply chain management, and maintenance. Factories will become fully automated with AI-driven robots performing complex tasks efficiently. Predictive maintenance, powered by AI, will ensure that machine downtime is minimized, leading to higher productivity and cost savings. The entire supply chain will be interconnected, utilizing real-time data to make smarter logistical decisions, reducing delays and optimizing resource allocation.

Agriculture stands to gain exponentially from the future of AIoT. Precision farming will see further enhancements with AI models predicting crop outcomes based on a myriad of variables like soil health, weather patterns, and pest activity. IoT devices will provide real-time data, allowing for timely interventions. Innovations such as autonomous farming equipment will become commonplace, leading to increased yield and sustainability. This technological leap will be crucial for feeding the world's growing population efficiently.

In the realm of energy and utilities, AIoT will play a pivotal role in guiding us toward a more sustainable future. Smart grids will become more intelligent, balancing energy loads in real-time and integrating renewable energy sources seamlessly. The predictive capabilities of AI will enable more efficient resource management, reducing waste and ensuring a more balanced supply-demand dynamic. Moreover, advancements in IoT-enabled infrastructure will facilitate the transition to renewable energy by optimizing the storage and distribution of energy generated from sources like solar and wind.

The retail industry will also be transformed by the future of AIoT. Customer experiences will be hyper-personalized, driven by AI analysis of consumer behavior and preferences. Smart shelves and inventory systems will ensure that stock levels are optimally managed, reducing waste and improving profit margins. The use of AI in logistics will make the supply chain more transparent, allowing for better tracking and delivery of goods. Virtual shopping assistants powered by AI will provide real-time recommendations, enhancing the shopping experience.

Transportation will undergo an incredible transformation through AIoT, with the rise of autonomous vehicles being just the beginning. AI-driven traffic management systems will optimize traffic flow in real-time, reducing congestion and emissions. Public transportation will be revolutionized with predictive maintenance and real-time updates, ensuring efficiency and reliability. Enhanced fleet management systems will improve logistics operations, making the delivery of goods faster and more cost-effective.

In the financial sector, AIoT will bring about unparalleled security and efficiency. Fraud detection systems will become more sophisticated, using AI to predict and prevent fraudulent transactions before they occur. Financial services will become more accessible and personalized, thanks to AI-driven customer service platforms. IoT devices will enable seamless financial transactions, integrating with smart home systems, wearable tech, and other connected devices to offer a truly omnichannel banking experience.

Entertainment and media will see the rise of intelligent, immersive experiences. AI will curate personalized content, transforming how we consume media, while IoT will enable interactive and connected viewing experiences. Smart home entertainment systems will integrate with various IoT devices, offering synchronized, multi-sensory experiences. Virtual and augmented realities driven by AI will allow for unprecedented levels of engagement, bringing media and entertainment into a new era.

From an ethical and privacy standpoint, the future of AIoT will require stringent regulations and robust frameworks to ensure data security and ethical practices. Ensuring the right balance between innovation and privacy will be critical. Companies will need to adopt ethical AI practices and enhance transparency to gain consumer trust. Meanwhile, governments and regulatory bodies will play a crucial role in shaping the guidelines that will safeguard individual privacy and ensure responsible use of AIoT technologies.

The roadmap ahead for AIoT is full of technological marvels that we can only begin to imagine today. Emerging trends like edge computing, 5G connectivity, and quantum computing will further augment the capabilities of AIoT, making devices smarter and more efficient. As these technologies converge, they will open up new frontiers and possibilities, pushing the limits of what can be achieved through AIoT integration.

In conclusion, the future of AIoT is not just an extension of current trends but a leap into a new era of intelligence and connectivity. We stand on the cusp of a technological revolution that will redefine every aspect of our lives, industries, and societies. Embracing this future with thoughtful consideration, ethical practices, and a commitment to innovation will ensure that the benefits of AIoT are realized for all. The journey may be challenging, but the destination promises a world that is smarter, more efficient, and profoundly interconnected.

CHAPTER 16:
AI AND IoT IN EDUCATION

The integration of AI and IoT in education is creating a more interactive, personalized, and efficient learning environment. With the advent of smart classrooms, students now have access to tailored educational content that adapts to their individual learning paces and styles, ensuring they receive the help they need exactly when they need it. AI-driven analytics offer insightful data to educators, allowing them to understand students' strengths and areas that require attention, leading to more effective teaching strategies. IoT devices, such as smartboards and connected lab equipment, enhance practical learning experiences, making complex subjects more accessible and engaging. Moreover, administrative operations benefit from automation, resulting in streamlined processes and better resource management, ultimately creating a more conducive environment for both teaching and learning. This seamless fusion of AI and IoT promises to transform traditional educational paradigms, preparing students with the skills they need for a tech-driven future.

Smart Classrooms

In today's rapidly evolving educational landscape, the concept of Smart Classrooms stands at the forefront of innovation. Integrating AI and IoT in educational settings is transforming the traditional classroom into an interactive, dynamic learning environment. Smart Classrooms are not just a theoretical idea; they're becoming a practical reality in many schools and universities worldwide.

At the heart of Smart Classrooms is the ability to create personalized learning experiences for students. AI algorithms analyze data gathered from various IoT-enabled devices, such as sensors, cameras, and smartboards, to understand individual students' learning patterns. This information is used to tailor educational content to meet the unique needs of each student. In essence, the use of technology has made it possible to move from a one-size-fits-all educational model to a more customized approach, significantly improving student engagement and learning outcomes.

Consider the role of smartboards in the modern classroom. These intelligent boards are equipped with AI capabilities that can recognize handwriting, convert written text to digital content, and even provide instant translations in multiple languages. This ensures that no student lags behind due to language barriers, making learning more inclusive and accessible.

Another cornerstone of Smart Classrooms is the use of AI-driven analytics to provide real-time feedback to both students and teachers. By evaluating performance data continuously, AI systems can identify areas where students might be struggling and suggest targeted interventions. Teachers, equipped with these insights, can provide more focused guidance, thereby fostering a more supportive learning environment.

IoT devices play a significant role in enhancing classroom management. Sensors can monitor environmental factors such as temperature, humidity, and lighting to create optimal learning conditions. Furthermore, smart seating arrangements powered by IoT can ensure equitable distribution of resources and attention, adapting seating plans based on students' requirements and performance data.

Security is another critical area where AI and IoT technologies are making a substantial impact. Smart Classrooms are equipped with surveillance cameras and access control systems that use facial recognition technology to ensure the safety of students and staff. This not only enhances security but also helps in maintaining attendance records more accurately and without manual intervention.

The concept of gamification is gaining traction in Smart Classrooms as well. AI-driven applications can incorporate game elements into the curriculum, making learning more engaging and fun. For instance, complex subjects like mathematics or science can be taught using interactive simulations and virtual labs, where students can experiment in a risk-free, guided environment.

Furthermore, Smart Classrooms facilitate collaborative learning. Interactive displays and connected devices allow students to work on group projects seamlessly, regardless of whether they are physically co-located or collaborating remotely. AI tools can provide real-time feedback on the group's dynamics, suggesting ways to improve collaboration and enhance the learning experience.

It's essential to consider the broader implications of Smart Classrooms on the educational ecosystem. Administrators can use data analytics to make informed decisions about resource allocation, curriculum development, and staff training. This enables a more efficient use of resources and helps in scaling educational programs to meet the growing demand.

Professional development for teachers is another crucial aspect. AI-powered training programs can provide personalized learning paths for educators, helping them to stay updated with the latest teaching methodologies and technological advancements. This not only enhances their skill set but also ensures that they can effectively use Smart Classroom technologies to improve student outcomes.

Though the benefits of Smart Classrooms are undoubtedly compelling, the adoption of these technologies is not without its challenges. Issues such as digital divide, data privacy, and the need for substantial investment in infrastructure can pose significant hurdles. However, the long-term advantages of creating a more personalized, efficient, and engaging educational environment make these challenges worth overcoming.

In conclusion, the integration of AI and IoT in Smart Classrooms is transforming education in unprecedented ways. By creating personalized, engaging, and secure learning environments, these technologies are paving the way for a brighter future in education. As we continue to explore and refine these innovations, the promise of Smart Classrooms offers a glimpse into the potential of what education could be: inclusive, efficient, and extraordinarily effective.

Personalized Learning

When it comes to education, one size does not fit all. Every learner is unique, and the integration of AI and IoT in personalized learning addresses this individuality by tailoring educational experiences to fit the needs, strengths, and preferences of each student. This isn't just a trend; it's a transformative innovation reshaping the entire educational landscape.

AI-driven algorithms mesh seamlessly with IoT devices to create dynamic learning environments that adapt in real-time. Imagine a classroom where sensors detect students' engagement levels, allowing AI systems to adjust teaching methods on the fly. This means that if a student is struggling with a maths problem, the system can instantly switch to a different explanatory approach, finding the path that resonates most. The era of static lesson plans is being replaced by dynamic, responsive teaching strategies.

The power of personalized learning lies in its capacity to *gather and analyze* vast amounts of data. AI algorithms scrutinize students' interactions, quiz results, and even their focusing patterns to build a comprehensive profile. These profiles aren't static; they evolve to reflect ongoing learning activities and outcomes, ensuring that educational content remains relevant and challenging.

It's not just about identifying weaknesses. AI and IoT technologies can also uncover hidden talents and interests that might otherwise go unnoticed in traditional classrooms. By exposing students to a wide array of subjects and monitoring their reactions, these systems can suggest new areas for exploration, providing a more rounded and enriched educational experience. A student excelling unexpectedly in an arts module might be directed towards additional resources, even suggesting extracurricular activities that align with their newfound passion.

The application of AI and IoT in personalized learning doesn't stop at content delivery. IoT-enabled devices such as smart desks and interactive whiteboards foster immersive learning environments. When combined with AI, these tools can provide instant feedback, gamify learning experiences, and promote collaborative projects that engage students more deeply than traditional methods ever could.

Take, for instance, an interactive whiteboard equipped with IoT sensors. The board might recognize when students are drawing or writing with hesitancy and prompt the AI system to offer hints or step-by-step guides. This kind of immediate, contextual assistance helps maintain the flow of learning and diminishes the frustration that can sometimes accompany complex topics.

Moreover, AI and IoT technologies offer the potential to democratize education. Personalized learning systems can operate at scale, making high-quality, customized education accessible to underserved communities. Imagine digital tutors that bridge gaps in educational resources, providing remote students with the same level of personal attention available to their urban counterparts.

Parental involvement also gains a new dimension through AI and IoT. Parents can receive timely updates on their child's progress, pinpointing areas of strength and those needing attention. Such insights foster informed conversations at home, aligning parental support with the educational strategies employed in classrooms.

The transition to personalized learning isn't without challenges. Privacy concerns regarding data collection and usage are paramount. Schools and technology providers must navigate these issues carefully, ensuring compliance with regulations and maintaining transparency with students and parents. Data security protocols have to be stringent to protect sensitive information from unauthorized access.

Additionally, teachers play a crucial role in leveraging these technologies effectively. Despite the sophistication of AI and IoT, human oversight remains invaluable. Educators need to be adept at interpreting AI-generated insights and integrating them into their teaching methodologies. Professional development programs focusing on technological fluency are essential to empower teachers to harness these tools fully.

Personalized learning also redefines the role of the student. Learners are no longer passive recipients but active participants in their educational journey. AI tools encourage self-paced learning, allowing students to delve deeper into topics that intrigue them while pacing through more familiar subjects at speed. This autonomy fosters a love for learning, as students navigate paths tailored to their interests and curiosities.

The future promises even more sophisticated systems as AI and IoT technologies continue to evolve. We may soon see AI that not only personalizes learning content but also predicts future educational needs, preparing learners for careers that haven't yet been conceived. Dynamic curriculum adjustments could ensure students are continually acquiring relevant skills, staying ahead in a rapidly changing job market.

In conclusion, the fusion of AI and IoT within the realm of personalized learning marks a revolutionary step in education. It promises to create a more inclusive, engaging, and effective educational environment where each student's potential is maximized. Empowering both teachers and students, these intelligent systems pave the way for a brighter educational future, customized and adaptive in ways previously unimagined.

Administrative Efficiency

Administrative efficiency in the education sector is undergoing a significant transformation with the integration of AI and IoT technologies. The days of manually handling attendance, scheduling, and record-keeping are gradually fading into the past, replaced by systems that can streamline processes, reduce errors, and free up valuable time for educators and administrators.

One fundamental aspect of administrative efficiency enhanced by AI and IoT is smart attendance tracking. Traditional methods, like roll calls and manual registers, are not only time-consuming but also prone to inaccuracies. IoT-enabled smart ID badges can automatically record student attendance as they enter and leave classrooms, while AI algorithms process this data in real-time. This system not only saves time but also guarantees accurate tracking, alleviating any discrepancies in attendance records.

Scheduling is another area where AI shines. Coordinating timetables for classes, exams, and even meetings can be a complex puzzle with countless variables. AI-driven scheduling software can analyze all the factors—teacher availability, room occupancy, student schedules—and optimize an arrangement that best fits everyone's needs.

This reduces conflicts and makes the entire educational environment more fluid and less prone to disruptions.

Data management and analytics represent another significant leap. Educational institutions generate massive amounts of data daily, from student grades and progress reports to administrative records. AI can automate the processing and analysis of this data, providing valuable insights. For instance, predictive analytics can identify at-risk students by analyzing patterns in attendance and performance, allowing for preemptive interventions.

Document and record management systems have also benefited from AI and IoT. Gone are the days of sifting through mountains of paper or outdated filing systems. Digital records, enhanced with AI algorithms, can be efficiently indexed and retrieved. This not only speeds up the administrative tasks but also ensures that records are accurately maintained and easily accessible when needed.

Moreover, the introduction of chatbots and virtual assistants into administrative roles has been a game-changer. These AI-driven entities can handle a myriad of repetitive tasks, such as answering frequently asked questions, booking appointments, and even grading simple assignments. By offloading these tasks onto AI, staff can focus more on strategic initiatives and less on monotonous duties, resulting in heightened efficiency.

Financial management within educational institutions is another avenue where AI and IoT have made considerable inroads. From managing budgets to optimizing resource allocation, AI can process financial records and provide actionable insights. Predictive algorithms can forecast future expenses, enabling institutions to plan more effectively and avoid potential financial pitfalls.

IoT devices also play a key role in maintaining physical infrastructure. Sensors installed in educational facilities can monitor everything from lighting and HVAC systems to security and safety protocols. These sensors collect data that can be analyzed in real-time to optimize energy usage, ensure safety compliance, and maintain a conducive learning environment. For example, smart lighting systems can adjust based on occupancy, reducing energy wastage and contributing to a more sustainable campus.

Communication channels within the institution are also enhanced. AI-powered communication platforms can facilitate seamless interaction between staff, students, and parents. Automated notifications and reminders ensure that everyone is on the same page, reducing the likelihood of missed deadlines and forgotten events.

In parallel, AI-enhanced plagiarism detection tools have revolutionized administrative processes related to academic integrity. These tools can swiftly scan student submissions against vast databases to identify potential instances of plagiarism, ensuring that academic standards are upheld without placing an undue burden on educators.

The benefits of AI and IoT in enhancing administrative efficiency aren't limited to K-12 institutions. Higher education and universities are also realizing immense advantages. From personalized advisor recommendations to course registration, AI streamlines numerous touchpoints, significantly reducing manual intervention and elevating the overall administrative capability of these institutions.

It's essential to recognize that while technology is a transformative tool, it doesn't operate in a vacuum. Institutional leadership and a proactive approach to change management are crucial. This includes training staff to proficiently use new systems and ensuring that the transition period is as smooth as possible. Proper implementation often requires a well-charted strategy that aligns with the institution's specific needs and goals.

While administrative tasks might seem behind-the-scenes compared to teaching and curriculum development, their efficiency has a cascading effect on the overall education system. When administrative processes are streamlined and effective, teachers can dedicate more time to educating, and students benefit from a more organized and responsive institution.

In summary, AI and IoT are pioneering the next phase of administrative efficiency in education. From automating mundane tasks to offering predictive insights, these technologies are reshaping how institutions operate, facilitating a smoother, smarter, and more efficient administrative environment. The result is an education system that not only functions better but also creates a more conducive atmosphere for learning and growth.

CHAPTER 17:
AIoT IN ENTERTAINMENT AND MEDIA

In the realm of entertainment and media, the fusion of AI and IoT, or AIoT, is creating a paradigm shift that's captivating audiences and transforming user experiences. AI-driven content creation and recommendation systems are reshaping how we discover and consume media, offering personalized suggestions tailored to individual tastes. Smart home entertainment systems, powered by interconnected IoT devices, are providing seamless, immersive viewing and listening experiences that adapt to user preferences and environmental contexts. Meanwhile, interactive media enriched with AI capabilities is paving the way for dynamic, real-time user engagement, blurring the lines between creators and consumers. The synergy of AI and IoT is not just enhancing entertainment—it's revolutionizing how stories are told, shared, and experienced, making the entire media landscape more intuitive, engaging, and responsive than ever before.

Content Creation and Recommendation

The integration of AI and IoT in the entertainment and media sectors has significantly reshaped the landscape of content creation and recommendation. This powerful amalgamation, often referred to as AIoT, leverages the capabilities of AI to process and analyze vast amounts of data, while IoT devices gather real-time information from multiple sources. The result is a more personalized, engaging, and efficient content ecosystem that continually evolves to meet the demands of its diverse audience.

189

AI and IoT together can redefine the way we consume and interact with content. Take, for instance, the role of AI algorithms in crafting personalized recommendation systems. These systems analyze viewers' behavior, preferences, and consumption patterns to suggest movies, TV shows, articles, and music that are most likely to interest them. Companies like Netflix, Spotify, and YouTube have revolutionized viewer experience using these technologies. They analyze vast datasets collected via IoT devices such as smart TVs, smartphones, and wearables to offer more accurate and timely content recommendations.

The depth and breadth of data that AI algorithms can process allow for nuanced understandings of audience preferences. This capability goes beyond simple genre suggestions. For instance, AI can analyze the thematic elements, pacing, and even the emotional tone of content to offer personalized recommendations that resonate on a more profound level. If a user shows a preference for fast-paced, high-action thrillers, the AI system can suggest similar content, even if it falls under a different genre but has a similar pace and structure.

Another aspect where AI and IoT play a critical role is content creation itself. AI-driven tools can assist creators by generating scripts, composing music, and even producing visual content. These tools analyze prevalent trends and audience reactions to help content creators understand what resonates most. By harnessing machine learning algorithms, writers can get assistance in drafting plots that are likely to engage viewers. This isn't about replacing the artist; rather, it enhances their creative process by providing insights that they might not have considered.

Moreover, AIoT can streamline the production process by automating routine tasks and optimizing workflows. IoT devices can monitor equipment, manage resources, and even control lighting and camera systems. Meanwhile, AI can continuously analyze data from these devices to predict potential issues and recommend adjustments on the fly. This leads to more efficient production cycles, reducing downtime and saving costs.

The rise of AIoT has also transformed live events and interactive media. IoT devices embedded in event spaces can collect data on audience engagement in real-time. AI analyzes this data to make on-the-spot adjustments to the content, like enhancing visuals or modifying the narrative flow to maintain audience interest. This dynamic interaction creates a more immersive experience that can evolve based on audience feedback, making each presentation unique and engaging.

Interactive media where the audience plays a part in shaping the content is another exciting development. Video games, virtual reality experiences, and interactive TV shows can now adapt their narratives and challenges based on player interaction and feedback collected via IoT devices. AI algorithms process this feedback to provide a more tailored and engaging experience. The result is a dynamic narrative that evolves in real-time, making the user feel more connected and invested.

Content recommendation also extends to advertisements and promotional content. AI-powered systems analyze user behavior and preferences to deliver targeted ads that are more likely to capture interest. The relevance and timing of these ads can significantly improve user experience, reducing the intrusion factor often associated with traditional advertising. By aligning ads more closely with user preferences, brands can achieve better ROI and build stronger relationships with their target audience.

AIoT's ability to track and analyze viewer engagement in real-time opens up new avenues for content monetization. For example, real-time data can help creators identify moments within content that generate high engagement, allowing them to place advertisements strategically. This method ensures ads are seen at peak interest, making them more effective. Additionally, real-time insights can help creators decide which content to promote heavily or adapt quickly based on viewer responses.

From the perspective of media platforms, AIoT-driven recommendation systems can significantly enhance user retention. By continually offering fresh, relevant, and engaging content, platforms can maintain higher viewer engagement and reduce churn rates. The continuous feedback loop created by IoT devices feeding data to AI systems ensures that the content evolves with viewer preferences, keeping the experience fresh and appealing.

Another fascinating use of AI in content creation is deepfake technology. While controversial, this technology enables the creation of hyper-realistic digital content, such as recreating historical figures or crafting new sequences for films without the need for physical actors. IoT can provide real-time motion capture, further enhancing the realism and fluidity of such digital creations. Ethical considerations aside, this technology showcases the potential of AIoT in pushing the boundaries of traditional content creation.

Voice-activated technologies powered by AI and connected through IoT devices are also contributing to the content landscape. Virtual assistants like Amazon Alexa and Google Assistant can suggest music, podcasts, or videos based on voice commands and previous interactions. Integrating these systems into smart homes means personalized content is just a voice command away, adding convenience and enhancing user experience.

The dynamic synergy between AI and IoT creates endless possibilities for innovation in content creation and recommendation. The ability to continuously learn from vast datasets, predict trends, and adapt to audience preferences in real-time gives creators and platforms an unprecedented edge. As technology evolves, we can expect even more personalized, engaging, and immersive content experiences that will redefine the boundaries of entertainment and media.

This transformative power of AIoT isn't just limited to large-scale players in the entertainment industry. Independent creators and small media companies can also leverage these technologies to level the playing field. With accessible AI-driven tools, they can analyze their audience's preferences, streamline their workflows, and offer highly personalized content. The democratization of these technologies means a more diverse array of voices and stories, enriching the content ecosystem as a whole.

Content creation and recommendation in the era of AIoT are on the cusp of something extraordinary. Imagine a future where content not only meets but anticipates our needs. It's a world where the boundary between creator and consumer blurs, enabling a more interactive and personalized media experience. This promise of continuous evolution, driven by data and enhanced by innovation, makes the future of AIoT in entertainment and media profoundly exciting. It's an era where each piece of content carries the potential to resonate on a deeply personal level, making every interaction meaningful.

Smart Home Entertainment Systems

In the current era of connected living, smart home entertainment systems are redefining how we consume media and interact with our entertainment devices. The seamless integration of AI and IoT within these systems has cultivated an experience that is not only immersive but also intuitive and personalized. Imagine walking into your living room and, without lifting a finger, your favorite show starts playing, the lighting adjusts for optimal viewing, and the temperature sets to your preferred comfort level. This futuristic scenario is no longer a distant dream but a present-day reality, thanks to advancements in AIoT technologies.

AIoT melds the decision-making capabilities of artificial intelligence with the connectivity and data-sharing functions of the Internet of Things. In smart home entertainment systems, this integration enhances both user convenience and system efficiency. AI algorithms analyze patterns in user behavior and preferences to offer personalized content recommendations, while IoT devices collect and share data to optimize system performance. The result is an entertainment ecosystem that feels extraordinarily responsive and attuned to the needs and desires of its users.

One of the standout features of smart home entertainment systems is their ability to offer personalized content recommendations. Traditional media consumption often required users to search or browse through vast libraries of content, a time-consuming and often frustrating process. AI algorithms leverage machine learning and data analytics to study user preferences, viewing habits, and even social media interactions to offer tailored recommendations. These suggestions are continually refined and updated, ensuring that users are always presented with content that aligns with their tastes and interests. This shift from passive to proactive content delivery transforms the way users discover and engage with media.

Voice-activated virtual assistants are another cornerstone of smart entertainment systems. Devices like Amazon Echo and Google Home have integrated seamless voice controls allowing users to interact with their entertainment setup without the need for physical remotes. Simply using a voice command, you can change the channel, control the volume, or even ask your AI assistant to find a specific movie or TV show. These assistants can also integrate with other smart home devices, enabling you to dim the lights or set the perfect mood for a movie night through a simple voice prompt.

However, it's not merely about convenience and personalization. The integration of AI and IoT also enhances the core technical capabilities of entertainment systems. For instance, AI can improve audio quality by dynamically adjusting settings based on room acoustics.

Similarly, video quality can be enhanced through AI-driven upscaling techniques which make standard definition content look almost as good as high definition. IoT devices, on the other hand, ensure that these systems are always running optimally, providing real-time diagnostics and predictive maintenance solutions to prevent potential issues before they escalate.

Interactive media experiences have also been revolutionized by AIoT. Gone are the days when entertainment was a one-way street. Nowadays, users can participate in interactive and immersive experiences that make them feel like a part of the story. Virtual reality (VR) and augmented reality (AR) technologies, powered by AI and IoT, offer hyper-engaging content that dynamically responds to user actions and inputs. These experiences can range from interactive movies to immersive gaming, providing myriad possibilities where consumers can lose themselves in entirely new worlds.

Privacy and security are paramount in smart home entertainment systems. With so many interconnected devices gathering and analyzing data, it's critical to ensure that this information is safeguarded against potential breaches. AI plays a crucial role in this aspect by offering robust security protocols and real-time threat detection mechanisms. Encryption and secure data transmission are standard features, ensuring that your entertainment experience is not only enjoyable but also safe.

Another fascinating application of AIoT in smart home entertainment systems is in the domain of energy management. These systems can monitor and analyze energy consumption patterns, optimizing device operations to reduce power usage without compromising performance. For instance, your smart TV might dim its brightness levels when it detects nobody is in the room, or the system might enter a low-power state during periods of inactivity. Such intelligent energy management features make the entire entertainment system not just smart but also eco-friendly.

Remote management and multi-device synchronization are significant advantages of these advanced systems. Through cloud-based platforms and mobile applications, users can control and manage their entertainment settings from virtually anywhere. Whether it's scheduling a recording while you're away, synchronizing audio across multiple rooms, or even monitoring usage statistics to better understand your viewing habits, AIoT provides the tools to maintain an uninterrupted and cohesive entertainment experience.

Moving forward, the evolution of smart home entertainment systems will likely be driven by further advancements in AI and IoT. As these technologies mature, we can expect even more sophisticated algorithms and smarter devices capable of delivering ultra-personalized experiences. Edge computing might play a significant role, reducing latency in real-time interactions and making systems faster and more efficient. Additionally, as 5G networks become more widespread, the connectivity and data-sharing capabilities of these systems will only enhance, leading to even more impressive and fluid interactions.

The future of smart home entertainment is incredibly promising, offering an ever-improving blend of convenience, personalization, and interactivity. These advancements are setting the stage for a new era of media consumption, where user experience is paramount, and technology seamlessly adapts to each individual's needs. The union of AI and IoT in this domain stands as a testament to what's possible when cutting-edge technologies converge, ultimately enriching our lives in ways we previously only imagined.

Interactive Media

In the vibrantly evolving landscape of entertainment and media, interactive media stands out as a groundbreaking frontier. These forms of media allow users not just to consume content passively, but to actively engage, modify, and even create content themselves. The rise of Artificial Intelligence (AI) and the Internet of Things (IoT) has catalyzed this transformation, enabling unprecedented levels of interactivity and personalization. This isn't just a niche within the entertainment world—interactive media is fundamentally reshaping how we experience stories, games, and even art.

Imagine, for instance, a video game that morphs its storyline based on your real-time decisions, or an art installation that changes in response to your emotions, captured through wearable sensors. This is the promise of AIoT in interactive media. By blending AI's cognitive capabilities with IoT's connectivity, creators can offer deeply personalized and immersive experiences. What was once science fiction is now within our grasp, enabling richer, more engaging forms of storytelling and interaction.

The power of AI in this context lies in its ability to analyze vast amounts of user data to predict preferences, make real-time adaptations, and create dynamic, evolving content. Meanwhile, IoT devices act as the sensory extensions of these intelligent systems, collecting data from the environment and from users themselves. From gesture control through wearable tech to voice commands captured by smart speakers, the interplay between these technologies forms the backbone of today's interactive media experiences.

Consider the role of AI-powered recommendation systems in streaming platforms. These algorithms analyze viewing patterns, user feedback, and even social media activity to suggest content that aligns precisely with individual tastes. This type of personalization not only enhances user satisfaction but also increases engagement and retention rates, forming a win-win for both consumers and service providers.

But the implications go beyond tailored movie picks. In augmented reality (AR) and virtual reality (VR), AI algorithms can generate virtual environments that adapt in real-time based on user movements and interactions. IoT sensors can capture not just where you are in a virtual space, but how you're feeling, translating these insights into dynamic adjustments within the environment. For example, a VR meditation app might alter its visual and auditory landscape to match the user's heart rate and stress levels, creating a perfectly tailored relaxation experience.

Interactive media also finds its place in social platforms. With AI and IoT, social experiences can become more engaging and meaningful. Imagine a virtual concert where the crowd's collective emotions, captured through smart wearables, influence the setlist or visual effects. These immersive experiences make users feel like active participants rather than passive spectators.

The role of AIoT in interactive media doesn't stop at entertainment. It has significant educational implications as well, creating opportunities for interactive learning experiences. AI can assess a student's progress in real-time and adapt the content to better suit their learning pace and style. IoT devices can provide sensory feedback, making the learning environment more interactive and engaging.

In the world of advertising, interactive media powered by AI and IoT is revolutionizing the way brands connect with consumers. Targeted advertising can be made not just based on browsing history but on real-time data collected from IoT devices. For instance, a smart refrigerator can suggest meal ideas based on the current inventory, paired with targeted ads for missing ingredients. This level of interactivity and personalization enhances the consumer experience while providing valuable insights for marketers.

Another exciting application is in the realm of live events. Imagine attending a live theater where the storyline can change based on audience reactions tracked in real-time, or a sports event where viewers can influence camera angles or access real-time statistics through connected devices. These innovations create a more immersive and engaging experience, bridging the gap between the physical and digital worlds.

Developing such sophisticated systems in interactive media involves tackling several technological challenges. Ensuring seamless integration between AI algorithms and IoT devices is paramount. This requires robust software engineering practices, extensive data collection, and rigorous testing to create a fluid user experience. Security and privacy are also critical considerations, as sensitive data is often at the core of these interactions. Developers must prioritize building systems that are not only intuitive but also secure and compliant with data protection regulations.

The democratization of content creation is another significant impact of AIoT in interactive media. Advanced tools powered by AI can assist even novice creators in producing high-quality content. Machine learning algorithms can handle complex tasks like video editing, sound mixing, and special effects generation, allowing creators to focus more on the creative aspects. IoT devices further enhance this process by providing real-time feedback and control, making content creation more accessible and efficient.

Looking ahead, the future of interactive media is bound to be even more transformative. Advances in AI and IoT, coupled with emerging technologies like 5G and edge computing, will push the boundaries of what's possible. We can anticipate experiences that are not only more immersive but also more intelligent, with systems that can learn and adapt over time to deliver ever more personalized and engaging content.

Interactive media fueled by AIoT is more than just a technological evolution; it's paving the way for a new paradigm in how we engage with content. It's fostering a world where the lines between creator and consumer blur, offering everyone a chance to be part of the narrative. This exciting convergence of AI and IoT is set to redefine the entertainment and media industries, creating richer, more interactive experiences that were previously unimaginable. The revolution is not just about the technology itself but about the incredible possibilities it opens up for creativity, engagement, and human connection.

CHAPTER 18:
ENVIRONMENTAL IMPACT

The integration of AI and IoT is revolutionizing the way we address environmental challenges. By enabling more precise climate change monitoring, these technologies provide real-time data that helps scientists and policymakers make informed decisions on emission reductions and environmental conservation. In waste management, AI-powered IoT systems optimize recycling processes and minimize landfill use, promoting a circular economy. Additionally, sustainable practices are elevated through smart resource management, allowing industries to reduce their carbon footprint and energy consumption. Embracing AIoT in environmental initiatives uncovers a future where technology and sustainability are synergistically aligned, paving the way for a healthier, greener planet.

Climate Change Monitoring

The advent of AI and IoT has revolutionized our capacity to monitor and address climate change. By leveraging these technologies, we can collect and analyze vast amounts of data that provide a clearer picture of our planet's health, enabling more effective decision-making and policy development. Climate change monitoring serves as a bedrock for any meaningful environmental strategy, ensuring that we are better prepared to mitigate its impacts and adapt to new realities.

AI excels at processing large datasets, identifying patterns, and forecasting future scenarios. When paired with IoT, which provides real-time data through various sensors and devices, the potential to monitor climate change with unprecedented precision becomes a reality. From satellite imagery to ground-based sensors, these integrated systems capture intricate details about the Earth's atmosphere, oceans, and land surfaces.

One of the primary applications of this technology is in tracking greenhouse gas emissions. Sensors embedded in industrial facilities, vehicles, and even natural environments can continuously monitor levels of carbon dioxide, methane, and other harmful gases. AI algorithms then analyze this data, providing insights into emission trends, identifying major contributors, and suggesting targeted interventions. This capability is pivotal for nations and organizations committed to reducing their carbon footprints and achieving sustainability goals.

Moreover, the collaboration between AI and IoT has empowered meteorologists and environmental scientists to predict extreme weather events with greater accuracy. The data collected from IoT devices, such as weather stations, drones, and satellites, is fed into AI models that simulate climate conditions and forecast events like hurricanes, floods, and heatwaves. Early warnings generated from these models can save lives and minimize economic losses by enabling timely evacuations and preparations.

Another crucial aspect of climate change monitoring is the assessment of deforestation and land degradation. IoT-enabled sensors and drones equipped with high-resolution cameras and LIDAR technology can survey vast tracts of land, detecting changes in forest cover and soil quality. AI processes this data to provide comprehensive reports on deforestation rates, endangered ecosystems, and areas requiring conservation efforts. This information is vital for governments, NGOs, and conservationists seeking to protect and restore natural habitats.

The health of our oceans is equally critical in the study of climate change. IoT-enabled buoys, underwater drones, and satellite imagery feed real-time data into AI algorithms that monitor sea level rise, ocean temperature, acidification, and marine biodiversity. This continuous monitoring helps scientists understand how climate change is affecting marine ecosystems and fish stocks, allowing for the development of adaptive management strategies to mitigate these impacts.

Furthermore, agricultural systems are increasingly benefiting from real-time climate data to enhance resilience against climate variability. IoT sensors monitor soil moisture, crop health, and local climate conditions, while AI analyzes this data to provide farmers with actionable insights. Predictive models offer recommendations on optimal planting times, irrigation schedules, and crop selection, reducing the risks posed by changing weather patterns and contributing to food security.

Urban areas are not left out either. Smart city initiatives integrate AI and IoT to monitor urban heat islands, air quality, and energy consumption. Sensors placed on buildings, streetlights, and transportation systems collect data that AI algorithms use to optimize energy usage, reduce heat generation, and improve air quality. This data-driven approach ensures that cities are not just adapting to climate change but actively participating in its mitigation.

Climate change monitoring also extends to global biodiversity, which is under threat due to shifting climate patterns. AI-powered cameras and IoT devices track wildlife movements, population dynamics, and habitat changes. This data helps conservationists devise and implement strategies to protect endangered species and maintain ecological balance. The detailed and continuous observation of wildlife interactions offers invaluable insights that static, intermittent studies cannot match.

Water resources management is another critical area benefiting from AI and IoT integration. Sensors monitor the quality, level, and flow of rivers, lakes, and reservoirs, while AI systems analyze this data to predict droughts, floods, and water contamination events. These

Charlie Morgan

insights are pivotal for water resource planning, ensuring that supplies remain stable and clean, even in the face of climatic disruptions.

Satellite-based IoT and AI collaborations have opened new frontiers in climate science. Satellite data offers a macroscopic view of climate phenomena, from polar ice melt to desertification. AI algorithms synthesize this information into user-friendly visualizations and predictive models, providing policymakers with clear, actionable insights. This bird's-eye perspective complements ground-based monitoring, creating a holistic picture of climate change.

The role of AI in climate modeling cannot be overstated. Traditional climate models require enormous computational power and time to simulate specific scenarios. AI accelerates this process by efficiently approximating the outcomes of complex climate models. These AI-driven models are continually refined with real-time data from IoT devices, making them increasingly accurate and reliable for long-term climate projections.

Additionally, public engagement and awareness are enhanced through AI and IoT. Informative apps and platforms that synthesize climate data from IoT devices help individuals understand their environmental impact. These platforms offer personalized advice on reducing carbon footprints, encouraging more sustainable behaviors. Spectacular visualizations derived from AI-analyzed IoT data capture the public's interest, fostering a culture of climate responsibility.

The importance of climate change monitoring cannot be emphasized enough. It's not just about collecting data; it's about transforming that data into actionable insights. AI and IoT together make this transformation possible, turning raw data into a powerful tool for climate action. The continuous flow of information and advanced analytics offers a dynamic, real-time picture of our environment, allowing us to respond more swiftly and effectively to the challenges posed by climate change.

In conclusion, the fusion of AI and IoT in climate change monitoring represents one of the most significant technological advancements in environmental science. This integration provides an intricately detailed and dynamic understanding of our planet's ever-changing conditions, enabling proactive measures to mitigate and adapt to climate impacts. Through continuous innovation and widespread adoption, AI and IoT are poised to be at the forefront of the global effort to safeguard our environment for future generations.

Waste Management

As our global population continues to rise, effective waste management has become more critical than ever. Traditional waste management methods, often reliant on labor-intensive processes and inefficient machinery, are increasingly inadequate. However, the integration of Artificial Intelligence (AI) and the Internet of Things (IoT) promises to revolutionize waste management, making it smarter, more efficient, and more environmentally friendly.

One of the primary ways AI and IoT transform waste management is through better monitoring and data collection. Smart sensors embedded in trash bins can collect real-time data on waste levels, types of waste, and frequency of disposal. This data is then transmitted to a central system where AI algorithms analyze it to optimize waste collection routes and schedules. The integration of these smart systems reduces unnecessary collections, resulting in lower fuel consumption and reduced carbon emissions. Essentially, it makes waste collection a lot more environmentally sustainable while cutting down operational costs for waste management companies.

Additionally, AI can bring advanced sorting methods to recycling centers. Historically, sorting recyclable materials from waste has been a laborious and often inaccurate process. By employing computer vision and machine learning algorithms, automated systems can now identify and separate different types of recyclable materials with high precision. These AI-powered systems can distinguish between types of plastics, metals, and papers, maximizing the amount of material that can be recycled and reducing the contamination rate in recycling streams.

Furthermore, IoT-enabled waste bins can communicate directly with both the collection authorities and the community. Imagine a smart city where waste levels in public bins are monitored in real-time and collection is automatically scheduled before the bins overflow. This would not only improve urban cleanliness but also prevent issues such as pests and bad odors, significantly enhancing the quality of life for the residents.

Embracing AIoT in waste management also opens the door to predictive analytics. Through continuous monitoring and data analysis, AI systems can predict waste generation patterns based on variables such as time of year, community events, and population changes. This predictive capability allows waste management companies to prepare better, allocate resources more efficiently, and even anticipate and manage waste surges during peak times.

The benefits extend to hazardous waste management as well. AI-powered systems can identify hazardous waste materials accurately and ensure they are handled with the appropriate safety measures. Sensors can detect hazardous substances that are not easily identifiable through traditional means, and AI can recommend the best ways to neutralize or dispose of these materials.

The integration of AI and IoT in waste management also supports the broader vision of a circular economy. By improving recycling rates and making waste processing more efficient, these technologies help ensure that resources are reused, reducing the need for raw material extraction. This shift to smarter waste management methods aligns well with global efforts to mitigate environmental impact and promote sustainable practices.

Moreover, the use of AI and IoT in waste management is instrumental in tackling specific environmental challenges like the reduction of plastic waste in oceans. Advanced tracking systems powered by GPS and IoT can monitor plastic waste paths from their point of origin to their final destination. Coupled with AI's predictive abilities, these systems can foresee potential pollution hotspots and enable timely interventions to collect the waste before it enters marine ecosystems.

Waste management also presents an opportunity for community engagement and education through AI and IoT. Imagine an AI-driven app that provides households with real-time feedback on their waste production and recycling habits. This app could offer tips on how to reduce waste and recycle more effectively, fostering a culture of environmental responsibility. Community initiatives can be further enhanced by gamification, where residents earn points for good waste management practices, invoking a sense of competition and community spirit.

The role of AIoT in waste management isn't confined to urban areas. Rural areas and developing regions, which often suffer from inadequate waste management infrastructure, can also benefit significantly. Portable IoT devices and drones equipped with sensors can be deployed to monitor waste management processes remotely. AI systems can analyze the data collected and provide actionable insights for improving waste handling practices in these regions, ensuring that the benefits of smarter waste management are universally accessible.

It's also worth noting the potential for job creation and industry growth as AI and IoT technologies proliferate in the waste management sector. While automation may reduce the need for certain manual roles, it creates new opportunities in tech development, system maintenance, and data analysis. Training programs and education initiatives focused on these new technologies can help workers transition to higher-skilled roles, contributing to economic growth and job security.

However, the integration of AI and IoT in waste management is not without its challenges. Data security and privacy concerns are paramount, especially when dealing with sensitive information. Effective frameworks must be in place to ensure that data collected by smart systems is protected and used responsibly. Moreover, there is the challenge of ensuring interoperability between diverse systems and devices, necessitating standardization and collaboration across industry stakeholders.

In summary, the fusion of AI and IoT is set to revolutionize waste management, making it more efficient, sustainable, and responsive to societal needs. It offers a roadmap for tackling environmental challenges while promoting economic and social well-being. By embracing these transformative technologies, we can move closer to a future where waste management is not just an afterthought but a cornerstone of sustainable urban living and environmental stewardship.

Sustainable practices

Sustainable practices in the context of AI and IoT integration are essential for ensuring that the rapid technological advancements of today do not come at the expense of our planet's future. These practices are not merely an afterthought but have become a central consideration as we continue to integrate smart technologies into various aspects of our daily lives and industries. From energy management to waste reduction, the convergence of AI and IoT offers transformative solutions that align with the goals of sustainability.

One of the most significant ways in which AI and IoT contribute to sustainable practices is through energy management. Smart grids, enhanced by AI algorithms and IoT sensors, can optimize the distribution and consumption of energy. These grids can predict and balance loads more efficiently, reduce waste, and integrate renewable energy sources like solar and wind more effectively into the power supply. This means fewer fossil fuel emissions and a lower carbon footprint.

The shift towards renewable energy sources is another area where AI and IoT provide substantial benefits. By leveraging predictive analytics and real-time monitoring, these technologies can enhance the efficiency and reliability of renewable energy systems. For instance, AI can forecast weather patterns to optimize the use of solar panels and wind turbines, while IoT devices can monitor equipment performance to preemptively address maintenance issues, thereby reducing downtime and maintaining consistent energy output.

In agriculture, sustainable practices are paramount for preserving natural resources and ensuring food security. Precision farming techniques, powered by AI and IoT, allow farmers to use resources more efficiently, applying water, fertilizers, and pesticides in the exact amounts needed. This not only minimizes waste and reduces harmful runoff into local ecosystems but also promotes healthier crop yields. Sensors in the soil and on plants can provide data that, when analyzed by AI, leads to actionable insights for farm management, promoting practices that not only improve productivity but also conserve resources.

Another crucial area is waste management. Traditional waste management systems are often reactive, dealing with waste once it has already become a problem. In contrast, AI and IoT can create proactive waste management strategies that reduce waste production from the outset. IoT-enabled bins that detect the type and volume of waste can communicate with waste management systems to optimize collection routes and recycling efforts. AI can then analyze this data to identify trends and recommend strategies for waste reduction at the source.

In the context of smart cities, sustainable practices involve a holistic approach to urban planning and operations. AI-driven data analytics can interpret vast amounts of data from various IoT devices across the city, aiding in the management of resources like water, electricity, and transportation. Smart lighting systems, for example, can adjust based on pedestrian presence and ambient light levels, reducing energy consumption. Public transportation systems can be optimized to reduce fuel usage and emissions, while also providing more efficient service to residents.

Water management is yet another critical aspect where sustainable practices are greatly enhanced by AI and IoT. Smart irrigation systems and leak detection technologies help conserve water by ensuring it is used only when necessary and detecting issues early on. AI can analyze weather and soil moisture data to optimize watering schedules, reducing water waste, and preserving this precious resource. This is particularly important in areas prone to drought or water scarcity.

The integration of AI and IoT also brings significant advancements in sustainable manufacturing practices. In industrial settings, predictive maintenance driven by AI can foresee machinery failures before they occur, minimizing downtime and the associated waste of resources. IoT sensors continually monitor machine health and performance, sending real-time data to AI systems that predict maintenance needs with high accuracy. This proactive approach ensures that manufacturing processes are as efficient as possible, reducing material waste and energy consumption.

Moreover, the move towards a circular economy is facilitated by AI and IoT technologies. A circular economy aims to keep resources in use for as long as possible, extracting maximum value before recovery and regeneration of products at the end of their lifecycle. AI can identify and optimize recycling processes, while IoT devices track and manage resources throughout their lifecycle. This reduces the need for raw materials and minimizes waste, promoting a more sustainable economic model.

Transportation, a major contributor to global greenhouse gas emissions, significantly benefits from AI and IoT in pursuing sustainable practices. Autonomous electric vehicles integrated with AI systems can optimize routes to reduce energy consumption, avoid traffic congestion, and minimize emissions. IoT-based traffic management systems can analyze and manage traffic flow in real-time, reducing idle times and associated pollution. Furthermore, shared mobility solutions powered by AI predict demand and optimize the use of transportation resources, promoting a more sustainable urban mobility ecosystem.

Sustainable practices are also evident in the realm of smart building management. Buildings are responsible for a large portion of energy use and emissions. AI can analyze data from IoT devices within buildings to optimize heating, ventilation, and air conditioning (HVAC) systems, lighting, and other operational parameters. This ensures that energy is used efficiently, reducing waste and enhancing the building's overall sustainability. Smart building technologies can also improve indoor air quality and comfort, creating healthier environments for occupants.

As AI and IoT technologies continue to evolve, their role in promoting sustainable practices becomes even more critical. The data generated by IoT devices, when combined with AI's analytical capabilities, provides deeper insights into how resources are used and where improvements can be made. This continuous feedback loop fosters a culture of sustainability, where practices can be adjusted and refined over time to achieve better outcomes.

Incorporating sustainable practices into AI and IoT integration isn't just beneficial for the environment—it also makes economic sense. Efficiency gains translate into cost savings, whether through reduced energy consumption, lower waste management expenses, or optimized resource usage. Companies that adopt these technologies can achieve substantial economic benefits while simultaneously fulfilling their corporate social responsibility goals.

Furthermore, government policies and incentives increasingly favor sustainable practices, providing additional motivation for industries to adopt AI and IoT solutions. Regulatory frameworks are evolving to support renewable energy adoption, waste reduction initiatives, and resource conservation efforts. Organizations that leverage AI and IoT to align with these frameworks can benefit from financial incentives, grants, or subsidies, further enhancing the viability and attractiveness of sustainable practices.

In conclusion, sustainable practices driven by the integration of AI and IoT are vital for addressing the environmental challenges of our time. By optimizing resource use, reducing waste, and enhancing efficiency across various sectors, these advanced technologies pave the way for a sustainable future. Embracing these practices not only benefits the environment but also drives economic growth and ensures that technological advancements enhance, rather than deplete, our planet's precious resources. As we continue to innovate, the principles of sustainability should remain at the forefront, guiding our path towards a more resilient and thriving world.

CHAPTER 19:
AIoT IN SMART HEALTHCARE

Integrating AI and IoT in healthcare is fundamentally transforming the way we understand and manage health. Through continuous remote health monitoring, AIoT enables real-time data collection from wearable devices, allowing for timely and precise interventions. AI-driven diagnostics leverage vast datasets to uncover intricate patterns that elude traditional methods, enhancing the speed and accuracy of diagnoses. Furthermore, IoT devices play a critical role in emergency response scenarios, providing immediate data to medical personnel and significantly improving patient outcomes. This fusion of AI with IoT is not only enhancing the efficiency of healthcare delivery but also personalizing patient care, driving us toward a future where proactive and preventive measures become the norm.

Remote Health Monitoring

Remote health monitoring stands as one of the most transformative applications of AIoT in smart healthcare. By integrating advanced AI algorithms with IoT devices, remote health monitoring leverages a seamless connection between patients and healthcare providers. This connection isn't just beneficial; it can be lifesaving. In a world where chronic diseases are on the rise and healthcare systems are increasingly burdened, remote monitoring offers a solution that enables continuous, real-time health data collection without the need for patients to constantly visit medical facilities.

At the heart of remote health monitoring are wearable devices and smart sensors that can track a variety of health parameters. These devices range from simple fitness trackers that measure steps and heart rate to sophisticated biosensors capable of monitoring blood glucose levels, blood pressure, and even complex cardiovascular metrics. These wearables are more than just gadgets; they serve as vital tools for both preventive and proactive healthcare management.

The data collected by these devices is vast and multidimensional, generating a comprehensive profile of a patient's health status. Here, AI plays a crucial role in processing this raw data. Through machine learning algorithms and predictive analytics, AI can identify patterns and anomalies that might escape the notice of human observers. For example, a subtle change in heart rate variability could be an early indicator of an impending cardiac event, allowing for timely intervention well before the situation becomes critical.

This constant stream of data is not only analyzed but also shared in real-time with healthcare providers. This connectivity is facilitated by the Internet of Things (IoT), which ensures that all the devices are synchronized and communicating effectively. The speed and efficiency with which this information flows can make the difference between a proactive healthcare measure and a delayed reaction to a medical emergency. Additionally, the Internet of Things creates a network of interconnected devices that can work collaboratively, offering a more holistic view of a patient's health state.

The benefits of remote health monitoring extend beyond mere data collection and analysis. For patients with chronic illnesses such as diabetes, hypertension, and congestive heart failure, continuous monitoring allows for better disease management. Patients can receive personalized care recommendations based on their unique health data, which is continually updated and refined by AI algorithms. This personalized approach can lead to improved health outcomes, reduce hospital admissions, and enhance the overall quality of life for patients.

Another significant advantage of remote health monitoring is its potential to address the issue of healthcare accessibility. In remote or underserved areas where medical facilities may be sparse, remote monitoring enables patients to receive the same level of care as those in urban centers. This democratization of healthcare means that geographical location no longer serves as a barrier to receiving high-quality medical attention. Thus, remote health monitoring contributes to narrowing the healthcare gap between different populations.

For healthcare professionals, remote health monitoring systems offer a wealth of data that can support better clinical decisions. Doctors and nurses can focus their attention on patients who exhibit signs of deteriorating health, guided by the insights provided by AI-powered analytics. This targeted approach allows for more efficient allocation of medical resources and can significantly mitigate the workload of healthcare providers. Additionally, it fosters a more proactive healthcare model, in which potential issues are identified and addressed before they become severe.

Equally important to consider is the impact of remote health monitoring on mental health. Stress, anxiety, and depression are often overlooked in traditional healthcare settings, yet they are critical components of overall well-being. Remote monitoring devices capable of tracking sleep patterns, mood fluctuations, and activity levels can provide valuable insights into a patient's mental health. AI algorithms can analyze these data points to offer mental health support and flag any concerning trends, prompting timely interventions.

Patients are not just passive recipients in this ecosystem; they become active participants in their health management. The integration of AIoT in remote health monitoring empowers patients with actionable insights and real-time feedback. They gain a deeper understanding of their health metrics and can engage in preventative behaviors based on AI-driven recommendations. This empowerment is motivating and can foster a more collaborative relationship between patients and healthcare providers, ultimately leading to better health outcomes and patient satisfaction.

While the promise of remote health monitoring is vast, it's not without challenges. Data security and privacy remain paramount concerns. The sensitive nature of health data necessitates robust cybersecurity measures to ensure that patient information is protected from potential breaches. Moreover, issues like data standardization and interoperability of devices must be addressed to create a seamless and effective remote monitoring ecosystem. Addressing these challenges will require concerted efforts from technology developers, healthcare providers, and regulatory bodies.

The role of artificial intelligence in interpreting the data is nothing short of revolutionary. Predictive analytics, machine learning models, and even neural networks could be employed to predict potential health issues. For instance, AIs trained on large datasets can predict the likelihood of stroke or heart attack based on a user's health metrics. These same algorithms can also be used to refine treatment plans, ensuring they're tailored to the individual in a way previously unimaginable. The ability of AI to make sense of complex data sets offers a strategic advantage in preemptive healthcare.

Additionally, remote monitoring can be a critical component during public health crises such as pandemics. The ability to monitor the health of large populations remotely can help in tracking the spread of diseases, identifying hotspots, and managing resource allocation. It provides a way to extend healthcare services without risking the safety of both patients and healthcare workers through unnecessary physical contact.

Moving forward, the integration of 5G technology promises to further enhance the capabilities of remote health monitoring. The high-speed, low-latency communication offered by 5G networks will allow for even more rapid data transfer and real-time analysis. It will enable more complex AI algorithms to be run on cloud platforms, enhancing the insights and predictive capabilities available to healthcare providers.

Equally important is the cost-effectiveness of remote health monitoring systems. The potential for reducing healthcare costs is significant, as these systems can minimize hospital admissions, reduce emergency room visits, and lead to more efficient use of medical resources. Insurance companies are starting to recognize these benefits, and some are beginning to offer incentives for using remote monitoring devices, which could further propel their adoption.

A broader societal benefit of remote health monitoring is the potential to contribute to large-scale health research. The aggregated data from millions of remote monitoring devices can offer unparalleled insights into public health trends, disease progression, and the effectiveness of treatment protocols. This treasure trove of data can help inform public health policies, healthcare practices, and future medical research, leading to advancements that can benefit the global population.

In conclusion, remote health monitoring represents one of the most promising advancements in the intersection of AI and IoT. By enabling continuous, real-time health tracking, these systems offer a proactive approach to healthcare that can significantly improve patient outcomes, enhance the efficiency of healthcare providers, and make quality healthcare accessible to even the most remote populations. The future of healthcare is undeniably leaning towards a more connected, data-driven approach, and remote health monitoring is at the forefront of this transformation.

AI-Driven Diagnosis

In the burgeoning frontier of AIoT in smart healthcare, AI-driven diagnosis stands prominently as a transformative force. The combination of artificial intelligence and Internet of Things (IoT) devices has initiated a paradigm shift in how diseases are identified and treated. This amalgamation promises not only to expedite diagnostic processes but also to increase their accuracy, all while significantly reducing human error.

At the heart of AI-driven diagnosis is the ability to analyze vast datasets with unprecedented speed and precision. Traditional methods of diagnosis often rely on a combination of patient history, physical examinations, and a limited set of diagnostic tests. In contrast, AI algorithms excel at processing large volumes of data from diverse sources, including electronic health records, medical imaging, and even wearable devices. These algorithms can identify patterns and anomalies that might elude the human eye.

For instance, machine learning models can be trained on thousands of radiographic images to distinguish between normal and pathological findings. Such models have demonstrated remarkable proficiency in detecting diseases like pneumonia, breast cancer, and diabetic retinopathy, frequently outperforming human radiologists. Furthermore, the iterative nature of these algorithms means that they continuously learn and improve over time, enhancing their diagnostic capabilities.

The integration of IoT devices plays a crucial role in gathering real-time health data. Wearable technology, such as smartwatches and fitness trackers, can monitor a patient's vital signs continuously. These devices transmit data such as heart rate, blood pressure, and oxygen levels to AI systems that analyze the information for signs of potential health issues. The immediacy and continuity of data collection inject a layer of dynamism into diagnostic processes, enabling early detection and intervention for conditions like arrhythmias or respiratory problems.

Moreover, AI-driven diagnostic systems can integrate various data streams to present a holistic view of a patient's health. This comprehensive approach allows for the analysis of interconnected factors that affect health outcomes. For example, an AI system might consider not only a patient's blood sugar levels but also their dietary habits, physical activity, and genetic predispositions to predict and diagnose conditions like diabetes with greater accuracy.

One real-world application of AI-driven diagnosis is in the field of cardiology. AI algorithms can analyze electrocardiograms (ECGs) to detect irregular heart rhythms suggestive of atrial fibrillation, a condition that can lead to stroke if left untreated. These systems can alert healthcare providers immediately, allowing for timely interventions that could be life-saving. In some cases, AI systems are employed to predict the risk of cardiac events based on the analysis of long-term data, thereby enabling preventative measures.

Oncology is another domain where AI-driven diagnosis is making significant strides. AI tools can evaluate pathologic slides and identify cancerous cells with high precision. Technologies like IBM's Watson for Oncology utilize natural language processing to parse medical literature and patient records, providing oncologists with evidence-based treatment recommendations. By offering superhuman-level data analysis, AI assists in formulating personalized treatment plans tailored to the specific characteristics of the patient's cancer.

A unique advantage of AI-driven diagnostic systems is their potential for democratizing healthcare. In remote and underserved regions, access to specialized healthcare providers is often limited. AI diagnostic tools, integrated with mobile health units or local clinics, can provide high-quality diagnostic capabilities to these communities. This deployment can significantly improve health outcomes by offering early and accurate diagnoses, which are critical in managing diseases effectively.

Despite its promise, the deployment of AI-driven diagnosis faces several challenges. One significant hurdle is the quality and variability of data. Accurate AI predictions depend on high-quality data for training and validation. Inconsistent data or biases present in the data can lead to inaccurate or misleading diagnostic results. Ensuring data quality and developing robust data governance frameworks are essential to overcoming this challenge.

Another consideration is the interpretability of AI models. Many AI systems operate as "black boxes," providing diagnostic results without explaining the underlying reasoning. This lack of transparency can hinder clinical adoption, as healthcare providers may be hesitant to rely on AI without understanding its decision-making process. Ongoing research in explainable AI aims to address this by making AI systems more interpretable and trustworthy.

Ethical and legal considerations also come into play. Issues such as patient privacy, data security, and informed consent are paramount. Using AI for diagnosis involves handling sensitive patient data, necessitating robust cybersecurity measures and stringent compliance with regulations like HIPAA. Ensuring patient consent and maintaining transparency about data usage are crucial in fostering trust and protecting patient rights.

Moreover, the integration of AI-driven diagnostic tools within clinical workflows requires careful planning. These tools should complement, rather than replace, human expertise. By integrating seamlessly into existing processes, AI can serve as a powerful aide to healthcare professionals, enhancing rather than diminishing their roles. Collaborative efforts, where AI augments the capabilities of healthcare providers, can lead to more efficient and accurate diagnosis and treatment pathways.

In conclusion, AI-driven diagnosis represents a revolutionary leap in healthcare. The fusion of AI analytics with IoT data collection offers unparalleled advances in diagnostic accuracy, speed, and accessibility. While challenges remain, the continual advancement of AI technologies and their ethical deployment can pave the way for a future where anyone, anywhere, can benefit from cutting-edge diagnostic care. By enhancing our ability to detect and respond to health issues early, AI-driven diagnosis not only improves patient outcomes but also builds a more resilient and responsive healthcare system. This transformative potential is only beginning to unfold, promising a healthier future for all.

IoT in Emergency Response

In the high-stakes world of emergency response, time is of the essence, and information is paramount. That's where the convergence of IoT (Internet of Things) and AI (Artificial Intelligence) technologies proves invaluable. The integration of IoT devices within emergency response frameworks has revolutionized how emergency situations are managed, providing real-time data and enhancing the responsiveness of systems. This section delves into the transformative impact of IoT on emergency response, illustrating how it enhances the core capabilities of healthcare systems and first responders, and ultimately saving lives.

Imagine a natural disaster scenario where a city is hit by a powerful earthquake. Traditional emergency response would depend heavily on human resources to evaluate the damage, coordinate rescue missions, and deliver medical aid. The advent of IoT changes this entirely. IoT-enabled devices can now collect and transmit critical data from disaster-stricken areas in real time. Smart sensors embedded in infrastructure can immediately report structural damages, while wearable health monitors can track the vital signs of survivors, sending alerts if immediate medical attention is needed.

Charlie Morgan

One significant application of IoT in emergency response is the deployment of smart ambulances. These vehicles are equipped with IoT devices that continuously monitor the health status of patients en route to the hospital. Parameters such as heart rate, blood pressure, and oxygen saturation are transmitted to the receiving medical facility in real time, allowing emergency rooms to prepare and prioritize treatments even before the patient arrives. This seamless flow of data can drastically reduce the time to treatment, which is critical in saving lives.

Furthermore, AI plays a crucial role in processing and analyzing the vast amounts of data generated by IoT devices. AI algorithms can sift through continuous streams of information from multiple sources, pinpointing patterns that human operators might miss. For example, AI can analyze traffic data to optimize ambulance routes, ensuring patients reach healthcare facilities as quickly as possible. This integration not only improves response times but also maximizes the effective use of resources.

IoT technology also extends its reach to smart firefighting systems. In the event of a fire, IoT-enabled sensors installed in buildings can detect smoke and heat, sending instantaneous alerts to emergency services. Moreover, these sensors can communicate with each other to map out the spread of the fire, aiding firefighters in strategizing their approach. Additionally, IoT can provide information on the presence of hazardous materials, helping in the formulation of safer firefighting strategies.

Another innovative development is the use of drones equipped with IoT devices for disaster management. These drones can access areas that may be dangerous or impossible for human responders to reach. Real-time video feeds and environmental data collected by drones can give emergency teams crucial insights into the situation on the ground, facilitating informed decision-making and effective coordination of rescue missions.

The realm of smart healthcare also benefits immensely from IoT in emergency response. For instance, during mass casualty incidents, IoT devices can be invaluable in triage situations. Wearable health monitors can automatically assess and grade the severity of injuries, transmitting this data to a central system where healthcare providers can quickly identify and prioritize the most critical cases. This technological assistance can streamline the triage process, ensuring that life-saving interventions are administered swiftly.

Of equal importance is the role of IoT in post-disaster recovery. IoT-enabled devices can monitor environmental conditions like air quality and the presence of contaminants, protecting recovery teams and displaced populations from potential health hazards. By providing real-time data on environmental risks, IoT contributes to safer and more efficient recovery operations.

Public health emergencies like pandemics also reveal the vital role that IoT and AI can play. IoT in conjunction with AI can help track and manage the spread of infectious diseases. Wearable health devices can monitor symptoms and vital signs, and when linked with AI, these devices can predict outbreak hotspots. This information equips public health authorities to implement targeted measures like quarantine and vaccination campaigns more effectively.

Moreover, IoT infrastructure extends beyond just data collection; it also facilitates communication and coordination among first responders and healthcare providers. Networked communication devices ensure that everyone involved in an emergency response is on the same page, reducing the chances of miscommunication and improving the overall efficiency of operations.

Innovation in IoT for emergency response is not confined to industrialized nations; it also holds tremendous potential for developing countries, which may lack robust healthcare infrastructure. Low-cost IoT sensors and mobile health devices can bring life-saving interventions to regions where immediate response capabilities are limited. This democratization of technology helps bridge the gap between different healthcare systems globally, ensuring better preparedness and response in all corners of the world.

Challenges remain in the implementation of IoT in emergency response, including issues related to data security, interoperability, and the need for robust network infrastructure. Protecting sensitive health data transmitted across IoT networks is of paramount importance. Additionally, ensuring that different IoT devices and systems can communicate seamlessly is essential for coordinated responses. Addressing these challenges will require ongoing innovation and collaboration among technologists, policymakers, and healthcare providers.

IoT in emergency response embodies a future where technology and human expertise work hand in hand to manage crises more effectively. The ability to gather, process, and act on real-time data brings a level of preparedness and agility that was previously unattainable. As IoT technology continues to evolve, it will undoubtedly open new avenues for enhancing emergency response, ultimately contributing to a safer, more resilient world.

Striving for innovation while adhering to ethical standards will be crucial as we advance. Safeguarding the integrity of collected data and ensuring its appropriate use will help maintain public trust in these powerful technologies. The seamless marriage of IoT and AI in emergency response is not just an aspirational concept; it is an achievable reality that stands to redefine our approach to managing emergencies, saving lives, and improving outcomes across the globe.

CHAPTER 20:
AI AND IOT SECURITY

As we continue our journey into the integration of AI and IoT, ensuring the security of these interconnected systems becomes paramount. The convergence of AI and IoT brings about immense benefits, but it also introduces new vulnerabilities and attack vectors that must be addressed to protect both data and devices. Cybersecurity measures play a crucial role in this landscape, where AI's sophisticated threat detection capabilities can proactively identify and neutralize potential risks before they escalate. Implementing robust security protocols to safeguard IoT devices is not only about preventing unauthorized access but also about maintaining trust in the technology that increasingly permeates our daily lives and industries. The resilience of AI and IoT systems against cyber threats will determine the extent to which we can harness their full potential, making security a foundational pillar for their successful deployment and continued innovation. By staying vigilant and adopting advanced security strategies, we can aspire to create a safe and secure technological environment where AI and IoT can thrive together.

Cybersecurity Measures

The rapid convergence of artificial intelligence (AI) and the Internet of Things (IoT), often referred to as AIoT, has unlocked unparalleled opportunities for innovation and efficiency across numerous sectors. However, this blend also introduces significant cybersecurity challenges that must be met with robust measures. Without adequate protection, the vulnerabilities inherent in interconnected devices and intelligent systems can be exploited, leading to catastrophic outcomes in both personal and industrial contexts.

First and foremost, understanding the fundamental vulnerabilities within AIoT systems is crucial. One prominent vulnerability arises from the sheer number of endpoint devices incorporated in an IoT network. Each device increases the potential attack surface, providing hackers numerous entry points to compromise the system. Securing these endpoints requires a multifaceted approach that includes device authentication, secure firmware updates, and rigorous access controls.

Device authentication serves as the first line of defense, ensuring that only authorized devices can access the network. In this regard, employing strong encryption protocols and multi-factor authentication provides a solid foundation for securing endpoints. That said, authentication on its own is insufficient. It must be complemented by continuous monitoring for anomalies that could indicate unauthorized access attempts.

Securing firmware updates is another critical aspect. Many IoT devices operate on outdated or unpatched software, making them prime targets for cyberattacks. Ensuring that devices can receive over-the-air updates in a secure manner is essential. This process involves digitally signing firmware packages so devices can verify authenticity before installation. Such measures prevent malicious software from gaining a foothold in the network infrastructure.

Access control mechanisms must be rigorously enforced to ensure that only authorized personnel and systems can interact with sensitive data and critical functions. Role-based access control (RBAC) is particularly effective, as it limits access permissions based on the user's role within the organization. In highly sensitive environments, attribute-based access control (ABAC) offers even finer granularity, incorporating user attributes and environmental conditions in the decision-making process.

Moving beyond individual device security, the network as a whole must be fortified against potential threats. Implementing network segmentation can isolate critical systems from less secure elements, thereby containing breaches and minimizing their impact. A segmented network architecture effectively limits the lateral movement of malicious actors within the system.

Furthermore, employing AI-based threat detection algorithms can significantly enhance the ability to identify and respond to cybersecurity threats in real-time. Machine learning models trained on vast datasets of known threats can recognize patterns indicative of cyberattacks, including zero-day exploits that traditional detection methods might miss. These models can then trigger automated responses, such as isolating affected segments and initiating forensic analysis.

Data security is equally paramount. As AIoT systems generate colossal amounts of data, ensuring its confidentiality, integrity, and availability becomes critical. Data encryption, both in transit and at rest, is essential to protect against unauthorized access and tampering. Additionally, implementing data anonymization techniques can help mitigate privacy risks, especially in applications involving sensitive personal information.

Intrusion detection systems (IDS) and intrusion prevention systems (IPS) play an essential role in maintaining network security. IDS tools continuously monitor network traffic for suspicious activities, providing alerts when potential threats are detected. IPS tools take an active role in blocking identified threats, often integrating with firewall technologies to provide a comprehensive security barrier.

One cannot overlook the importance of a robust incident response plan. Despite the best preventive measures, breaches can still occur.

Having a well-defined incident response plan enables organizations to react swiftly to mitigate damage, conduct thorough investigations, and restore systems to a secure state. This plan should encompass roles and responsibilities, communication protocols, and procedures for both containment and recovery.

Collaboration and information sharing among organizations are also pivotal in the fight against cyber threats. Participating in industry-specific cybersecurity forums and information-sharing consortia can provide valuable insights into emerging threats and best practices. Sharing threat intelligence allows organizations to proactively adapt their defenses based on the latest attack vectors observed across the industry.

Regulatory compliance adds an additional layer of complexity but is non-negotiable. Various global and regional regulations, such as the General Data Protection Regulation (GDPR) in Europe and the California Consumer Privacy Act (CCPA) in the United States, impose stringent requirements on data handling and cybersecurity practices. Ensuring compliance with these regulations not only safeguards against legal repercussions but also enhances overall security posture.

Furthermore, the role of human factors in cybersecurity can't be underestimated. Many cybersecurity breaches result from human error, such as falling victim to phishing attacks or neglecting to follow security protocols. Conducting regular cybersecurity awareness training for all employees is vital. This training should include information on recognizing phishing attempts, the importance of strong passwords, and the critical role each individual plays in maintaining security.

Adopting a "Zero Trust" approach marks a significant shift in cybersecurity philosophy. Zero Trust operates on the principle that threats could originate both outside and within the network. Therefore, it involves a continuous process of verifying the trustworthiness of every entity, whether it's a user, device, or application. By continuously validating access permissions and monitoring all activities for signs of compromise, Zero Trust mitigates the risks associated with an overly permissive security posture.

In summary, the cybersecurity measures needed to protect AIoT systems are expansive and multifaceted, encompassing device security, network protection, data integrity, and human factors. By leveraging advanced technologies such as AI for threat detection and adopting best practices in access control and incident response, entities can robustly defend against the ever-evolving landscape of cyber threats. As AI and IoT continue to redefine industries and enhance everyday life, robust cybersecurity measures will be the cornerstone of safe and effective integration.

AI in Threat Detection

As the integration of AI and IoT continues to expand, the issue of security has become more critical than ever. With countless devices interconnected through the IoT network, there's a heightened risk of cyber threats. Fortunately, the application of AI in threat detection is revolutionizing the field of cybersecurity, offering real-time solutions that can identify and neutralize risks before they cause harm.

AI-driven threat detection systems are designed to continuously monitor network traffic and device activity, leveraging machine learning algorithms to identify patterns and anomalies. These systems can differentiate between normal behavior and potential threats, whether they're malware, unauthorized access attempts, or data breaches. By learning from historical data, AI systems can predict future threats and preemptively tackle them, making networks more resilient.

One of the standout features of AI in threat detection is its ability to process and analyze vast amounts of data at speeds unattainable by human analysts. Traditional cybersecurity measures often rely on signature-based detection, which requires constant updates and can only identify known threats. In contrast, AI uses heuristics and anomaly detection to uncover unknown or zero-day threats, providing a superior level of protection.

A key component of AI in threat detection is the use of deep learning neural networks. These advanced models can analyze complex datasets, such as those from IoT devices, to identify subtle patterns that could indicate a security threat. By simulating the way the human brain processes information, deep learning models can understand the context and relevance of captured data, enabling more accurate threat identification.

Utilizing AI in threat detection isn't just about spotting and stopping attacks. It's also about hardening the security posture of IoT devices and networks. Machine learning algorithms can evaluate the vulnerability of various devices, suggesting updates and patches that minimize the attack surface. This proactive approach ensures that devices remain secure, even as new vulnerabilities emerge.

Moreover, AI can automate many of the routine tasks associated with cybersecurity management. For instance, AI systems can automatically deploy security patches, update firewalls, and manage encryption keys. Automation not only reduces the workload on IT teams but also ensures that these critical tasks are performed consistently and accurately.

The integration of AI and IoT also fosters a more holistic approach to threat detection. AI systems can aggregate data from multiple sources, including IoT sensors, network logs, and endpoint devices. This comprehensive view enables a more thorough analysis of the entire network environment, identifying threats that might otherwise go unnoticed.

Behavioral analysis is another crucial AI capability in threat detection. Rather than relying solely on predefined threat signatures, AI systems can learn the normal behavior of users and devices within the network. Any deviation from this baseline can trigger alerts, prompting further investigation. This method is particularly effective in identifying insider threats and sophisticated attacks that evade traditional detection methods.

The use of AI in threat detection also extends to defending against Distributed Denial of Service (DDoS) attacks, which aim to overwhelm network resources. AI systems can detect and mitigate DDoS attacks by analyzing traffic patterns to differentiate between legitimate and malicious requests. By dynamically allocating resources and blocking offending IP addresses, AI can ensure the continuous availability of services.

One of the challenges in deploying AI-driven threat detection systems is ensuring that they are transparent and interpretable. While AI offers powerful capabilities, it can also be seen as a 'black box' - complex and difficult to understand. Efforts are being made to develop explainable AI models that provide insights into how decisions are made, which helps build trust and facilitates compliance with regulatory requirements.

There's also the issue of false positives, where legitimate activities are flagged as threats. AI systems must be fine-tuned to balance sensitivity with specificity, reducing the likelihood of false positives without compromising security. Continuous learning mechanisms allow these systems to improve over time, refining their detection capabilities.

Real-world applications of AI in threat detection are numerous and varied. In the banking sector, AI systems monitor transactions for signs of fraud, alerting security teams to suspicious activities. In healthcare, AI safeguards patient data against unauthorized access, ensuring compliance with privacy regulations. These examples showcase how AI's versatility enhances security across different industries.

Collaboration is also key in the realm of AI and IoT security. Enterprises are working together, sharing threat intelligence to bolster their defenses against common threats. Open-source platforms and consortiums enable the pooling of resources and knowledge, advancing the development of more robust AI-driven security solutions.

While AI in threat detection offers immense benefits, it's essential to recognize that it's not a silver bullet. Security is a multi-faceted issue that requires a comprehensive strategy, encompassing technology, policies, and human vigilance. AI should be viewed as an augmentative tool that enhances, rather than replaces, traditional security measures.

Looking to the future, the role of AI in threat detection will only become more prominent. Advances in quantum computing, edge AI, and other emerging technologies promise even greater capabilities in defending IoT networks. The ongoing evolution of AI algorithms will continue to push the boundaries of what's possible, ensuring that security keeps pace with the rapid growth of IoT.

In summary, AI in threat detection is a transformative force in the domain of IoT security. By harnessing the power of machine learning and deep learning, AI systems can detect and neutralize threats with unprecedented accuracy and speed. As the IoT landscape evolves, these intelligent systems will be instrumental in safeguarding our interconnected world, enabling innovations and enhancing our quality of life.

Protecting IoT Devices

In a world increasingly dominated by interconnected devices, safeguarding IoT (Internet of Things) devices from a multitude of security threats has become paramount. With the integration of Artificial Intelligence (AI) and IoT, the complexity and potential vulnerabilities of these systems have multiplied. The vast network of smart home appliances, industrial machinery, healthcare devices, and transportation systems creates a sprawling attack surface that cybercriminals are more than eager to exploit. Consequently, it's essential to implement stringent security measures tailored to protect these IoT devices from evolving threats.

First and foremost, securing IoT devices begins with robust authentication mechanisms. Traditional password protection is often inadequate, especially given the limited computing capabilities of many IoT devices. Instead, multi-factor authentication (MFA) and unique device identification techniques are instrumental in enhancing security. MFA requires users to provide multiple forms of verification, such as a password coupled with a biometric scan or a mobile device confirmation, which significantly raises the bar for malicious actors attempting unauthorized access.

Encryption is another critical component in protecting IoT devices. Data transmitted between devices and over networks should be encrypted to prevent interception and manipulation by attackers. Advanced Encryption Standard (AES) and Transport Layer Security (TLS) protocols can be employed to ensure data integrity and confidentiality. For devices with limited processing power, lightweight encryption algorithms designed specifically for IoT environments, such as SPECK and SIMON, can be utilized to maintain data security without overburdening the device.

Beyond encryption and authentication, continuous monitoring and anomaly detection are essential in identifying and mitigating potential threats. AI systems proficient in machine learning can be deployed to monitor IoT devices and their communication patterns in real-time. By analyzing vast amounts of data, these AI models can detect anomalies indicative of potential security breaches. For instance, an unusual spike in network traffic or abnormal behavior patterns can trigger alerts, allowing for swift investigation and response.

Network segmentation can also play a vital role in securing IoT devices. By dividing a network into isolated segments, it becomes more challenging for attackers to move laterally and access sensitive systems. Segmenting IoT devices from critical business systems and other sensitive data repositories is a proactive approach to limit the scope of a potential breach. Each segment can be governed by stringent access controls and security policies to further reduce the risk of unauthorized access.

The principle of least privilege should be uniformly applied across IoT systems. This principle dictates that devices, users, and applications are granted only the minimum level of access necessary to perform their functions. By restricting excess permissions, the impact of a compromised device or application can be contained, preventing attackers from gaining broader access to the network.

Regular software updates and patch management are non-negotiable elements in the quest to protect IoT devices. IoT devices often run on firmware that can contain vulnerabilities discovered post-deployment. Manufacturers must be diligent in releasing firmware updates to address these security flaws. Equally important is ensuring that these updates are applied promptly. Automated update mechanisms can help maintain the security posture of IoT devices by ensuring they are always running the latest, most secure versions of their software.

User awareness and education should not be overlooked in protecting IoT devices. Many vulnerabilities stem from user practices, such as the use of default credentials or failure to update devices. Educating users about the importance of changing default settings, creating strong passwords, and remaining vigilant about firmware updates can significantly enhance the security of IoT environments.

Device manufacturers have a crucial role to play in embedding security from the design phase itself. Security by design ensures that security considerations are integrated into every element of the device, from hardware to software architecture. This proactive approach includes secure boot processes, hardware root of trust, and built-in encryption modules, making the devices resilient against various attack vectors.

Moreover, leveraging blockchain technology has shown promise in enhancing the security of IoT devices. Blockchain's immutable nature can ensure data integrity and secure transactions across a decentralized IoT network. Smart contracts on the blockchain can automate and enforce security policies, adding an additional layer of protection.

Interoperability and standards compliance are other vital considerations. Given the diversity of IoT devices and the varying protocols they operate on, adhering to industry standards such as those set by the Internet Engineering Task Force (IETF) and the Institute of Electrical and Electronics Engineers (IEEE) can ensure a baseline level of security. Standardized protocols facilitate easier and more secure integration of devices from different manufacturers.

One of the often understated aspects of IoT security is the consideration of physical security. Many IoT devices are deployed in locations that are not under constant supervision, making them susceptible to physical tampering. Installing devices in secure enclosures or using tamper-evident designs can augment their security. Coupled with frequent physical inspections, these measures can help detect and prevent unauthorized physical access.

Finally, a comprehensive incident response plan tailored for IoT environments is indispensable. In the event of a security breach, having a well-defined plan that includes steps for immediate isolation of compromised devices, thorough investigation, damage assessment, and swift recovery can mitigate the impact. Regular drills and updates to the incident response plan ensure preparedness to handle various types of threats.

Charlie Morgan

The journey of securing IoT devices is ongoing and requires a multi-faceted approach, combining advanced technologies, user awareness, and industry-wide collaboration. As AI continues to advance and integrate deeply with IoT, the landscape of threats will also evolve. However, by employing robust and adaptive security measures, we can build resilient IoT ecosystems that not only enhance our quality of life but also safeguard our digital future.

Chapter 21:
Regulatory and Legal Aspects

As AI and IoT technologies continue to grow and evolve, navigating the regulatory and legal landscape becomes crucial for their successful integration into everyday life and industry. Governments and regulatory bodies are working diligently to establish comprehensive frameworks that balance innovation with public safety and privacy. Current regulations aim to address data protection, ethical AI practices, and the interoperability of devices. However, rapid technological advancements often outpace policy development, leading to legal ambiguities that stakeholders must navigate. Yet, these challenges present opportunities for proactive engagement and collaboration between lawmakers, industry leaders, and technologists to craft policies that foster innovation while safeguarding societal interests. The ongoing dialogue in regulatory circles ensures that the deployment of AIoT solutions remains both forward-thinking and responsible, setting the stage for a future where technology enhances the human experience holistically.

Current Regulations

The landscape of regulations governing AI and IoT is continuously evolving, driven by rapid technological advancements and the increasing integration of these technologies into daily life and industry. Regulatory bodies across the globe are striving to keep pace with the myriad implications of AI and IoT, aiming to ensure that innovation can flourish while safeguarding public interests. The balance between promoting technological development and protecting the public from potential risks is delicate, but crucial.

In the United States, the federal government has adopted a somewhat decentralized approach to AI and IoT regulation. Agencies such as the Federal Trade Commission (FTC), the Federal Communications Commission (FCC), and the National Institute of Standards and Technology (NIST) each play roles in overseeing different aspects of these technologies. The FTC, for instance, focuses on consumer protection issues, including data privacy and cybersecurity, while the FCC handles wireless communication standards critical for IoT devices. Meanwhile, NIST develops standards and guidelines to foster innovation and industrial competitiveness in AI and IoT sectors.

One noteworthy piece of legislation is the European Union's General Data Protection Regulation (GDPR), which significantly impacts companies operating within or interacting with the EU. GDPR emphasizes data protection, granting individuals greater control over their personal data and imposing stringent requirements on its processing and storage. This regulation affects AI and IoT by ensuring that systems incorporating these technologies maintain high privacy standards. GDPR's reach extends beyond Europe, influencing global companies to adopt compliant practices to avoid hefty fines and penalties.

Asia is also actively developing frameworks for AI and IoT, with countries like China and Japan leading the charge. China's "New Generation Artificial Intelligence Development Plan" outlines the nation's strategy to become the world leader in AI by 2030. This ambitious plan includes regulatory measures to oversee AI development and deployment, focusing on both innovation and ethical considerations. Japan, similarly, has implemented its "Society 5.0" initiative, integrating AI and IoT into its societal infrastructure, with regulations ensuring safety and privacy in these emerging technologies.

Beyond national regulations, international bodies are also stepping in to create a harmonized regulatory environment. Organizations such as the International Telecommunication Union (ITU) and the Institute of Electrical and Electronics Engineers (IEEE) are working on standards that transcend borders. These efforts aim to foster global collaboration and ensure that AI and IoT systems can interoperate seamlessly while adhering to shared safety and ethical guidelines.

Despite these initiatives, regulatory challenges abound. The fast-paced nature of AI and IoT innovation often outstrips the ability of legislative processes to respond effectively. Policymakers must navigate complex technical landscapes, often lacking the expertise required to fully understand the implications of these technologies. This lag can lead to gaps in regulation, where advances in AI and IoT outpace the creation of appropriate safeguards, potentially exposing society to unforeseen risks.

Moreover, the ethical dimensions of AI and IoT present unique regulatory challenges. Issues such as algorithmic bias, surveillance, and the potential for job displacement require nuanced approaches that balance innovation with societal well-being. Ethical guidelines, like those from the European Commission's High-Level Expert Group on Artificial Intelligence, provide frameworks for responsible AI use but need to be continually updated to reflect new developments and insights.

In the commercial sphere, companies are often at the forefront of regulatory adaptation, implementing compliance measures to align with existing laws and anticipating future regulations. Businesses need to integrate regulatory compliance into their operational strategies, ensuring that their AI and IoT deployments meet both current and prospective legal requirements. This proactive approach not only minimizes legal risks but also builds consumer trust, which is crucial for the widespread adoption of these technologies.

Collaboration between stakeholders—governments, industry, academia, and civil society—is essential for effective regulation. Such partnerships can help create informed policies that strike the right balance between fostering innovation and protecting public interests. Initiatives like public consultations and multi-stakeholder forums allow for diverse perspectives, contributing to more robust and adaptive regulatory frameworks.

Looking ahead, it is clear that the regulatory environment for AI and IoT will need to remain flexible and agile. As technologies continue to evolve, so too must the regulations that govern them. Forward-thinking policies that anticipate future trends and challenges will be key to ensuring that the benefits of AI and IoT can be fully realized while mitigating potential harms.

Policy Development

As Artificial Intelligence (AI) and the Internet of Things (IoT) meld more firmly into the fabric of various industries and everyday life, crafting robust policies becomes not just necessary, but inevitable. Regulatory frameworks must evolve in tandem to accommodate the new paradigms brought forth by AIoT. This task, however, is intricate and multi-faceted, involving various stakeholders from governments, industry experts, and the public. Policy development acts as a crucial linchpin that connects technology with society, ensuring that innovation is channeled towards positive and sustainable outcomes without compromising public welfare and safety.

Effective policies are essential to foster innovation while mitigating risks. They shouldn't only offer guidelines for the ethical and secure use of AIoT applications but should also stimulate an environment conducive to technological advancement. Policymakers must address critical aspects such as data privacy, cybersecurity, and the ethical implications of automation. These regulations need to be dynamic, flexible, and forward-thinking, reflecting the rapid pace at which technology evolves.

One fundamental issue in policy development is data privacy. AIoT applications generate vast amounts of data, creating unprecedented opportunities for insights but also raising significant privacy concerns. Policymakers must develop regulations that safeguard personal information without stifling innovation. Transparent data handling protocols and robust encryption standards are essential components of any policy framework. Regulations like the General Data Protection Regulation (GDPR) in Europe have set a precedent, emphasizing the necessity for clear, enforceable standards. However, a one-size-fits-all approach may not be practical globally due to varying cultural and legal landscapes.

Cybersecurity is another cornerstone in AIoT policy development. With a myriad of connected devices communicating over networks, the potential for cyber threats increases exponentially. Governments and organizations must collaborate to establish stringent cybersecurity measures. Policies should mandate regular updates and patches for software, robust authentication mechanisms, and strict access controls. In addition, international cooperation is vital for tackling cross-border cyber threats, necessitating coherent and harmonized international policies.

Ethical considerations form a crucial part of the policy landscape for AIoT. As AI algorithms become integral to decision-making in critical areas like healthcare, finance, and autonomous transportation, ensuring these systems operate ethically is paramount. Policymakers need to establish frameworks that prevent biases, promote transparency, and ensure AI decisions are understandable and explainable. This will require ongoing dialogue between technologists, ethicists, and legislators to continually refine and adapt policies as the technology advances.

The rapid pace of AIoT development presents unique challenges for policymakers. The traditional regulatory process, often lengthy and slow, must adapt to keep up with technological innovations. Agile policy-making, characterized by iterative development, stakeholder involvement, and rapid prototyping, could be the answer. Experimental regulatory sandboxes, where new technologies can be tested in a controlled environment, allow for real-world insights into how policies might affect technological deployment and adoption. This approach encourages innovations while ensuring that potential risks are identified and managed early.

Furthermore, stakeholder engagement is essential in policy development. Policymakers should involve a broad spectrum of stakeholders, including technology developers, industry leaders, academia, civil society, and end-users. This inclusive approach ensures that multiple perspectives are considered, resulting in well-rounded policies that address the concerns of all parties involved. Public consultations and feedback mechanisms can facilitate this engagement, making the policy development process more transparent and democratic.

International collaboration is another critical element. With AIoT applications often transcending national borders, a coherent global framework is essential. Organizations like the United Nations, the International Telecommunication Union (ITU), and the World Economic Forum are already fostering international dialogues on AI and IoT governance. Countries need to work together to develop standardized regulations that facilitate cross-border cooperation and data exchange while maintaining national security and privacy standards.

Funding and resources also play a crucial role in the successful implementation of AIoT policies. Governments should allocate adequate resources for regulatory bodies to effectively monitor and enforce these policies. Additionally, investments in research and development are essential to keep pace with technological advancements. Public-private partnerships and international collaborations can help pool resources and expertise, promoting innovation and adherence to regulatory standards.

Education and training are indispensable for the development and adoption of effective AIoT policies. Policymakers, regulators, and the general public need to be well-informed about AIoT technologies, their potential benefits, and associated risks. Educational initiatives, workshops, and training programs can help bridge the knowledge gap, fostering a greater understanding and more informed decision-making. Furthermore, incorporating AIoT-related topics into educational curriculums can prepare the next generation for a future dominated by these technologies.

Finally, continuous monitoring and evaluation of policies are essential. The AIoT landscape is dynamic, and policies need to evolve accordingly. Regulatory bodies should establish mechanisms for the ongoing assessment of policy efficacy, identifying areas for improvement and making necessary adjustments. Feedback from stakeholders and real-world case studies can provide valuable insights, ensuring that policies remain relevant and effective in the face of ongoing technological advancements.

In summary, the development of policies governing AI and IoT is a multi-dimensional task that requires a delicate balance between fostering innovation and ensuring public safety. By addressing key areas such as data privacy, cybersecurity, and ethics, and embracing agile policy-making, international collaboration, and stakeholder engagement, we can create a conducive environment for the sustainable growth of AIoT technologies. With the proper frameworks in place, AIoT has the potential to revolutionize industries and enhance everyday life, driving us towards a future where technology serves humanity in unprecedented ways.

Legal Challenges

The convergence of AI and IoT introduces a landscape brimming with potential and complexity, especially in terms of legal challenges. These innovations are pushing the boundaries of what our current legal frameworks can handle, necessitating both adaptation and novel interpretations of existing laws. This section addresses the multifaceted legal issues that arise from integrating AI and IoT, focusing on liability, intellectual property, data privacy, and regulatory compliance.

One of the most significant legal challenges is the question of liability. Who is responsible if an AI-driven IoT device causes harm or malfunctions? Traditional notions of liability are centered on human accountability, but AI-driven systems often operate with a degree of autonomy that complicates this paradigm. When an autonomous vehicle crashes, or a smart home system fails, the involved parties could range from software developers, hardware manufacturers, and even data providers. Establishing a clear legal responsibility in such scenarios is not straightforward and requires new frameworks for attributing liability.

Intellectual property issues also present a considerable challenge. The integration of AI into IoT devices often involves sophisticated algorithms and large datasets that are proprietary. Determining who owns the rights to these algorithms, especially when they are improved or adapted by the machine itself, is a complicated legal matter. Moreover, the collaboration between multiple stakeholders in developing AIoT solutions further muddies the waters of intellectual property ownership. Protecting these innovations while promoting an open market for technological advancement demands a delicate balance.

Data privacy is another critical legal challenge in the realm of AI and IoT. These technologies thrive on data—lots of it. Everything from personal health information collected by wearable devices to location data tracked by smart city infrastructure raises issues of privacy and consent. Existing laws like the General Data Protection Regulation (GDPR) in Europe and the California Consumer Privacy Act (CCPA) in the United States provide some safeguards. However, they were not specifically designed with the capabilities and intricacies of AIoT in mind. Ensuring compliance with existing regulations while also updating these frameworks to address AIoT-specific nuances is essential for the technology's sustainable development.

Regulatory compliance itself poses a significant challenge. The rapid pace of AI and IoT innovations often outstrips the speed at which new regulations can be formulated and enforced. Regulatory bodies worldwide are struggling to keep up with the rapid technological advancements, and as a result, many existing laws are either outdated or inadequate. Regulatory uncertainty can stifle innovation, as companies may be hesitant to invest in new technologies without clear guidelines. Conversely, overly strict or poorly conceived regulations can also hamper technological progress. Thus, finding the right regulatory balance is a continuous challenge that requires collaboration between industry stakeholders and policymakers.

Another aspect of legal challenges in AIoT is cross-jurisdictional issues. Data and devices powered by AI and IoT don't respect geographical boundaries. Data collected in one country could be processed in another and stored in a third part of the world. This global flow of data presents significant legal hurdles in terms of compliance with various national and international laws. Different countries have different regulations concerning data privacy, cybersecurity, and even intellectual property. Navigating this complex web of jurisdictional differences is a daunting task for organizations operating on an international scale.

Moreover, the use of AI in making decisions brings up the challenge of algorithmic accountability. AI algorithms that inform IoT devices need to be transparent and fair to maintain user trust and legal compliance. Issues related to how these algorithms make decisions—whether they discriminate based on race, gender, or other factors—are under increased scrutiny. Ensuring that these algorithms meet ethical standards while still being effective is both a technical and legal challenge. Laws may need to be introduced or updated to ensure fairness and accountability in the deployment of AI technologies.

The pace of innovation in the AI and IoT space also means that legal frameworks need to be highly adaptive. Emerging technologies such as blockchain, which can offer solutions to some AIoT challenges like data security and transparency, also demand new regulatory considerations. The intertwining of these cutting-edge technologies creates a layered legal environment that requires proactive and futuristic regulatory measures.

Legal challenges in AIoT also extend to the realm of international collaboration and standardization. As countries and industries develop their own standards and regulations, there is a danger of fragmentation, which can impede the global adoption and interoperability of AIoT solutions. International cooperation and the development of universally accepted standards are necessary to ensure that AIoT technologies can integrate smoothly across borders.

Lastly, there's the issue of updating existing legal principles to encompass AI and IoT capabilities. Many current laws were drafted with human actors in mind and might not apply seamlessly to autonomous systems. Tort law, contract law, and consumer protection laws, among others, will need to be revisited and possibly overhauled to provide clear rules and protections in an AIoT-driven world.

Addressing these legal challenges is not merely a task for legislators and lawyers; it requires a concerted effort from technologists, ethicists, and policymakers. By understanding the multifaceted legal implications of AI and IoT, stakeholders can develop more robust, fair, and effective laws and regulations that will guide the responsible evolution of these transformative technologies.

CHAPTER 22:
BUSINESS STRATEGIES FOR AIoT

Stepping into the realm of AIoT, businesses must craft strategies that harness the full potential of this powerful convergence. A thorough market analysis is the first move, ensuring that companies understand current trends, customer needs, and potential growth areas. Armed with this insight, competitive strategies can be implemented to differentiate offerings, focusing on unique value propositions and leveraging technological advancements. Furthermore, investment in AIoT should be approached with a keen eye on ROI; this means not only initial costs but also long-term benefits and scalability. Strategic partnerships, robust R&D programs, and a customer-centric approach are essential components. By effectively aligning these elements, businesses can not only remain competitive but also lead the charge in the ever-evolving landscape of AIoT.

Market Analysis

Understanding the market dynamics for AIoT (Artificial Intelligence of Things) is essential for businesses looking to harness its transformative potential. AIoT represents the fusion of AI capabilities with IoT devices, offering enhanced analytics, improved efficiency, and more intelligent decision-making. As we delve into a detailed market analysis, we'll discover the various segments driving growth, the key players involved, and the future trends that could shape the industry.

Initially, the AIoT market has been divided into several segments, each reflecting a different application area. From healthcare to smart cities, the diversity of use cases allows for a broad range of market opportunities. For instance, the healthcare sector is making strides in remote monitoring and diagnostics, where AI algorithms analyze data from IoT devices to offer real-time health insights. On the other hand, industrial sectors benefit from predictive maintenance, where AIoT systems preempt machinery failures, reducing downtime and operational costs.

One can't ignore the market drivers stimulating the growth of AIoT. The increasing adoption of advanced technologies such as cloud computing, big data, and 5G networks has created a conducive ecosystem for AIoT solutions. The deployment of 5G, in particular, enables faster and more reliable connectivity, crucial for real-time data processing and analysis. Moreover, as businesses aim to optimize operations and reduce costs, the demand for AIoT solutions that provide actionable insights continues to rise. This trend is evident across various industry verticals, from manufacturing and retail to energy and utilities.

Key market players are shaping the AIoT landscape through constant innovation and strategic partnerships. Companies like IBM, Microsoft, and Siemens are at the forefront, offering integrated solutions that combine AI algorithms with IoT platforms. Startups are also contributing significantly, often specializing in niche markets or novel applications. These collaborations and investments help in pushing the boundaries of what AIoT can achieve, whether it's through enhanced software capabilities or the development of new sensor technologies.

In examining the geographical outlook, North America currently dominates the AIoT market, followed closely by Europe and the Asia-Pacific region. The presence of major tech companies, coupled with robust technological infrastructure, gives North America an edge.

Meanwhile, the rapid industrialization and digitization efforts in countries like China and India are propelling the Asia-Pacific region to become a significant player in the near future.

Moreover, regulatory environments play a substantial role in market development. Countries with favorable regulations and incentives for IoT deployment and AI research are likely to see accelerated growth in their AIoT markets. Policymakers are increasingly aware of the potential benefits, and this awareness is translating into supportive frameworks and investments into smart initiatives. Conversely, stringent data privacy laws and security concerns can act as hurdles, necessitating strategies that balance innovation with compliance.

On the horizon, emerging trends offer a glimpse into the future of the AIoT market. The integration of edge computing, where processing occurs closer to data sources, is gaining traction. This reduces latency and enhances real-time decision-making capabilities, crucial for applications like autonomous driving and smart grids. Additionally, advancements in AI models, including deep learning and reinforcement learning, promise more sophisticated and accurate analytics, further expanding the scope of AIoT solutions.

Sustainability is another trend influencing market dynamics. As environmental concerns gain prominence, AIoT solutions geared towards optimizing energy consumption, minimizing waste, and promoting sustainable practices are garnering attention. These solutions not only meet regulatory requirements but also align with the broader corporate agenda of social responsibility.

It's also worth noting how economic factors can influence the market. Economic downturns may lead to reduced capital expenditure, affecting the rate of new AIoT deployments. However, the push for cost-efficiency and productivity gains during such periods can also drive the adoption of AIoT, as businesses seek to do more with less.

Lastly, consumer behavior is another critical factor. The increasing consumer insistence on smart, connected devices—be it wearables, smart home systems, or connected cars—signals a growing market. The IoT devices in consumer hands generate immense volumes of data, opening up endless possibilities for AI-driven services and applications.

In summary, the market analysis for AIoT showcases a vibrant, evolving landscape brimming with opportunities. Businesses are encouraged to stay agile, leveraging these insights to remain competitive and successful in their AIoT ventures. The future promises a proliferation of smart solutions making daily life and industrial processes more interconnected and intelligent, paving the way for a smarter world.

Competitive Strategies

In the highly dynamic landscape of AI and IoT (Internet of Things), developing robust competitive strategies is paramount for businesses seeking to establish and maintain a foothold. The fusion of these two transformative technologies into AIoT presents unique opportunities and challenges. To navigate this environment successfully, businesses must adopt strategic approaches that set them apart from the competition while simultaneously fostering innovation and growth.

One fundamental aspect of competitive strategy in AIoT is differentiation. Differentiation can be achieved through unique product offerings, advanced technological capabilities, or exceptional customer service. For instance, a company that integrates cutting-edge AI algorithms with IoT sensors to provide real-time predictive maintenance solutions can distinguish itself from competitors. By offering a product that not only enhances operational efficiency but also reduces downtime, such a business can create significant value for its customers, thereby establishing a competitive edge.

Another crucial element is agility and adaptability. The AIoT space is characterized by rapid technological advancements and shifting market trends. Companies that can swiftly adapt to these changes by continuously innovating their products and services will likely stay ahead of the competition. This might involve regular updates to software and systems to incorporate the latest AI advancements or expanding IoT networks to cover new applications and industries.

Furthermore, cost leadership can also be a potent competitive strategy in the AIoT domain. By optimizing production processes and leveraging economies of scale, businesses can reduce costs and offer their products at more competitive prices. For instance, an AIoT company might develop a streamlined manufacturing process that lowers the cost of IoT devices without compromising on quality. This cost advantage can then be passed on to customers, making the company's offerings more attractive in a price-sensitive market.

Strategic partnerships and alliances play a pivotal role in enhancing competitive positioning within the AIoT industry. Collaborations with key technology providers, research institutions, or industry leaders can provide access to new technologies, expertise, and market channels. For example, a partnership between an AI-driven analytics company and a major IoT sensor manufacturer can result in the development of highly integrated solutions that offer superior performance compared to standalone products. These alliances can also facilitate the sharing of best practices and resources, driving innovation and efficiency.

It is also essential for businesses to focus on scalability. As the deployment of AIoT solutions grows, the ability to scale operations efficiently becomes a critical competitive factor. Developing versatile and modular solutions that can be easily scaled up to meet increasing demand or adapted to new applications can provide a strategic advantage. Additionally, cloud-based platforms and edge computing technologies can be leveraged to ensure that AIoT solutions remain robust and responsive as they scale, thereby maintaining high performance and customer satisfaction.

Investing in research and development (R&D) is another vital component of a competitive strategy for AIoT businesses. Constantly pushing the boundaries of what is possible through AI and IoT can lead to groundbreaking innovations and new applications. R&D investments not only fuel product development but also help companies stay abreast of emerging trends and technologies. Businesses that prioritize R&D are more likely to develop proprietary technologies and intellectual property that can serve as formidable barriers to entry for competitors.

Market analysis and understanding consumer needs are critical for developing effective competitive strategies. Analyzing market trends, customer preferences, and competitor activities provides valuable insights that can inform product development and marketing strategies. Businesses can use advanced AI-driven analytics tools to gather and interpret large volumes of data, enabling them to anticipate market shifts and customer demands more accurately. This proactive approach ensures that businesses can quickly adjust their strategies to align with market realities.

Providing exceptional customer support and service is another strategy that can significantly enhance competitiveness in the AIoT market. AI-powered customer service solutions, such as chatbots and virtual assistants, can deliver prompt and personalized support, ensuring high levels of customer satisfaction. Moreover, IoT-enabled devices can offer remote diagnostics and troubleshooting capabilities, allowing for quick resolution of issues without the need for on-site technicians. These customer-centric approaches not only build loyalty but also enhance the overall user experience, making the company's offerings more appealing.

Security and privacy considerations are increasingly becoming a competitive differentiator in the AIoT space. Given the growing concerns over data breaches and cyber threats, companies that prioritize robust security measures and transparent privacy practices are likely to gain the trust and confidence of their customers. Implementing advanced AI-driven cybersecurity solutions that protect IoT devices and networks from potential threats can give businesses a competitive edge. Furthermore, adhering to stringent data privacy regulations and being transparent about data usage can reinforce a company's reputation as a responsible and trustworthy player in the AIoT market.

Finally, businesses must also consider the long-term sustainability of their competitive strategies. Sustainable practices not only contribute to environmental conservation but also resonate with increasingly eco-conscious consumers. Companies that integrate sustainability into their AIoT solutions—such as optimizing energy usage through smart grids or developing eco-friendly IoT devices—can position themselves as leaders in responsible innovation. This forward-thinking approach not only differentiates them from competitors but also aligns with global trends toward sustainability and responsible business practices.

In conclusion, the competitive landscape of AIoT is complex and multifaceted. Businesses must adopt a combination of differentiation, agility, cost leadership, strategic partnerships, scalability, R&D investment, market analysis, customer support, security focus, and sustainability to thrive. By crafting and implementing comprehensive and adaptive competitive strategies, companies can not only succeed in the AIoT market but also drive the next wave of technological innovation.

Investment and ROI

Investing in AIoT requires a nuanced understanding of both the technologies involved and the potential benefits they bring. With an ever-growing market, businesses must evaluate the strategic merit of committing resources to AIoT. The integration of Artificial Intelligence (AI) with the Internet of Things (IoT) is creating a symbiotic relationship that promises substantial returns but comes with its set of challenges. The objective is not merely to implement AIoT for the sake of technological advancement; the end goal is to ensure a favorable Return on Investment (ROI).

ROI in the context of AIoT must be measured on multiple fronts. Financial returns are usually the primary focus, but other dimensions such as operational efficiency, customer satisfaction, and competitive advantage are equally important. Businesses are beginning to realize that the benefits of AIoT extend far beyond mere cost savings. Improved decision-making, enhanced predictive maintenance, and optimized supply chain operations are just a few examples of how AIoT can create value.

One compelling reason to invest in AIoT is the insights it offers through advanced data analytics. By leveraging AI algorithms, companies can sift through the vast amounts of data generated by IoT devices. This is crucial for identifying trends, forecasting demand, and making data-driven decisions. Consider the healthcare sector, where AIoT is being used for predictive diagnostics and real-time patient monitoring. The ROI here is not just financial but also extends to saving lives and improving patient care.

For many industries, the initial cost of implementing AIoT solutions can be daunting. These costs include hardware, software, integration services, and employee training. However, the long-term benefits often outweigh these initial expenses. Consider the manufacturing industry, where AIoT-driven predictive maintenance can significantly reduce downtime and extend the lifespan of machinery. Such improvements can lead to substantial cost savings and thus contribute positively to ROI.

The ROI of AIoT can also be seen through enhanced customer experiences. In the retail sector, for example, AIoT can offer personalized shopping experiences by analyzing consumer behavior and preferences. Stores equipped with AIoT solutions can manage inventory more efficiently, ensuring that popular items are always in stock while minimizing waste. The convenience and satisfaction for the customer translate to higher sales and brand loyalty, thereby boosting ROI.

Data security, while often seen as a challenge, is another area where investment in AIoT can yield returns. Robust AI algorithms can monitor IoT networks for unusual activities, thereby enhancing cybersecurity measures. The reduction in data breaches and associated costs can be substantial, leading to a more secure and resilient business operation.

Another significant aspect of ROI is the competitive edge AIoT can provide. Early adopters of AIoT technologies often set industry standards and become market leaders. The ability to innovate and offer solutions that competitors can't match is invaluable. For instance, in the energy sector, companies using AIoT for efficient resource management and smart grids can not only reduce operational costs but also attract environmentally conscious consumers.

However, achieving a favorable ROI from AIoT investments is not automatic; it requires strategic planning and execution. First, businesses need to identify specific areas where AIoT can provide the most significant impact. A clear understanding of business objectives, coupled with a thorough market analysis, can guide these decisions. Employing pilot projects can help test the waters before full-scale deployment, thereby mitigating risks.

In addition, the scalability of AIoT solutions is crucial for improving ROI. Solutions must be designed to scale seamlessly as business needs grow. Scalable solutions allow companies to start small and expand gradually, ensuring steady improvements in ROI over time. For this, cloud-based platforms often offer a flexible and scalable infrastructure, making it easier for businesses to adapt to increasing demands.

Customization and adaptability are vital as well. While off-the-shelf AIoT solutions offer a quick start, they may not meet specific industry needs. Custom solutions, though potentially more expensive initially, can provide higher ROI by addressing unique business challenges more effectively. This bespoke approach ensures that every dollar invested contributes directly to achieving targeted outcomes.

Training and development of the workforce involved in AIoT projects should not be overlooked. Skilled employees are required to implement, manage, and optimize AIoT systems effectively. Investing in training programs enhances the ability to utilize AIoT tools to their full potential, thereby improving efficiency and increasing ROI.

Collaboration with technology partners can also contribute to higher ROI. Engaging with specialized firms or startups can bring expertise and fresh perspectives that enhance the effectiveness of AIoT initiatives. Partnerships can also spread the financial risk, making substantial investments more feasible.

Maintaining a focus on continual innovation ensures sustained ROI. AIoT is a rapidly evolving field, and what works today might be outdated tomorrow. Businesses must stay abreast of technological advancements and incorporate new features and improvements regularly. This proactive approach helps maintain a competitive edge and ensures that the investment continues to yield benefits in the long run.

A critical component in calculating ROI is time-to-value. Understanding how quickly the investment will start to pay off is essential for planning and allocation of resources. Quick wins can help build momentum and secure further buy-in from stakeholders. Hence, deploying AIoT solutions in phases can help demonstrate early successes and validate the investment.

Lastly, measuring the ROI of AIoT investments requires appropriate metrics and KPIs. Financial metrics like cost savings, revenue growth, and profit margins are straightforward. However, other metrics such as customer satisfaction scores, uptime percentages, and efficiency ratios can provide a more comprehensive view of the impact. Developing a robust framework for monitoring and evaluating these metrics ensures that the ROI is transparent and quantifiable.

The integration of AI and IoT represents a paradigm shift in how businesses operate and compete. While the initial investment might appear substantial, the potential returns in terms of efficiency, customer satisfaction, and competitive advantage are significant. Businesses that strategically plan and execute their AIoT initiatives, while continually innovating and adapting, stand to gain immensely. The promise of AIoT is not just in the technology itself but in the transformative potential it holds for creating sustainable and scalable value.

CHAPTER 23:
IMPLEMENTING AIoT SOLUTIONS

Integrating AIoT solutions into any setup requires meticulous planning and design to ensure that both AI and IoT components work coherently towards a unified goal. Deployment strategies must be robust, adaptable, and scalable to accommodate future technological advancements and evolving user needs. Real-world implementations often highlight the tremendous potential and diverse applications of AIoT, acting as invaluable case studies for new projects. Organizations should focus on creating a seamless blend of AI-driven analytics with IoT's data-gathering prowess, thereby unlocking unprecedented efficiencies and innovative capabilities. This synergistic approach can transform industries, elevate user experiences, and drive unprecedented growth by leveraging intelligent, interconnected systems. Ultimately, effective implementation not only demands technical expertise but also a strategic vision aligned with the organization's goals and the specific challenges of the operational environment.

Planning and Design

The journey of implementing AIoT solutions begins long before deploying hardware or coding algorithms. It starts at the drawing board where meticulous planning and innovative design lay the groundwork for success. As we delve into this critical phase, we uncover the labyrinthine process that ensures the seamless integration of AI and IoT—an endeavor that calls for a harmonious alignment of technological prowess, strategic foresight, and creative ingenuity.

Planning and designing AIoT solutions is akin to setting up the architecture for a sprawling metropolis. Each element needs to be thoughtfully envisioned, interconnected, and scalable. The primary goal isn't just to solve a problem but to create a transformative ecosystem that can adapt and grow over time. This involves laying down a robust framework that ensures interoperability, security, and reliability right from the outset.

A significant starting point in the planning phase is needs assessment. Understanding the specific requirements of the industry or application at hand is paramount. Whether it's enhancing precision farming, revolutionizing healthcare, or optimizing smart grids, the needs assessment helps define clear, actionable objectives. This phase also involves engaging with stakeholders to gather insights and expectations, ensuring the solution delivers real value.

Once the needs are comprehensively understood, the next step is to define the architecture. The AIoT architecture encompasses hardware, software, network capabilities, and data flow. At this stage, particular emphasis is placed on choosing the right sensors and devices, setting up a reliable network infrastructure, and ensuring compatibility between different system components. Selecting the appropriate IoT platform and AI technologies that can handle vast amounts of disparate data in real time is crucial.

With the blueprint of the system architecture in place, the focus shifts to data strategy and management. Data is the lifeblood of AIoT solutions. Thus, defining how data will be collected, stored, processed, and analyzed is a foundational element of the design process. Implementing a data governance framework ensures data quality, security, and compliance with regulations. Architecting data flow from edge devices to the cloud, and ensuring real-time processing capabilities, becomes a pivotal consideration.

Security by design is an axiom that can't be overstated. Protecting AIoT networks against cyber threats requires a multi-layered security approach integrated into every layer of the architecture. This includes securing endpoint devices, ensuring encrypted data transmission, implementing robust access control measures, and continuously monitoring for anomalies. Designing a system with built-in security protocols sets a solid foundation for safeguarding sensitive information and maintaining user trust.

Another critical aspect of planning and designing AIoT solutions is scalability. Systems should be designed to handle the anticipated growth in the number of connected devices and data volume. This involves leveraging cloud services and edge computing to ensure seamless scalability and latency reduction. A modular design, where additional components or capabilities can be integrated without a complete redesign, ensures future-proofing the solution.

The user experience (UX) should be a focal point in the design phase. Effective UX design ensures that the interface is intuitive, user-friendly, and meets the end-users' needs. Involving UX designers early in the planning process helps in creating interfaces that simplify complex functionalities and drive user engagement. Conducting usability testing and gathering user feedback iteratively can refine the design and enhance the overall experience.

Interoperability is yet another key element in the design phase. The AIoT ecosystem comprises various devices, platforms, and technologies that need to communicate seamlessly. Employing open standards and protocols can facilitate interoperability, enabling diverse systems to work together cohesively and efficiently. Ensuring that the solution can integrate with existing infrastructure and future technologies is vital for long-term success.

Moreover, sustainability should be integral to the planning and design of AIoT solutions. Designing energy-efficient systems that minimize resource consumption not only reduces operational costs but also contributes to overall environmental goals. Implementing energy-saving protocols and leveraging renewable energy sources where possible is a step towards creating more sustainable and responsible AIoT solutions.

Another crucial factor during this phase is regulatory compliance. Navigating the regulatory landscape ensures the AIoT solution adheres to industry standards and legal requirements. Understanding data protection laws, industry-specific regulations, and upcoming policy changes allows for designing compliant and future-ready systems. This also involves establishing protocols for regular audits and adherence to compliance norms.

Testing and validation are fundamental steps as the solution blueprint transitions from paper to prototype. Rigorous testing under real-world conditions, stress tests, and validating AI models against diverse datasets ensure the system performs reliably and meets expectations. A robust validation process helps identify potential issues and iteratively improve the design before full-scale deployment.

Additionally, effective project management ensures the planning and design phases stay on track and within budget. Coordinating between cross-functional teams—including developers, designers, engineers, and stakeholders—facilitates smoother execution and timely updates. Utilizing agile methodologies can enhance flexibility, responsiveness to changing requirements, and faster delivery cycles.

The planning and design phase isn't just about creating a technical solution; it's about envisioning a holistic ecosystem ready to drive innovation and efficiency. It bridges the gap between conceptual ideas and tangible implementation, setting the stage for revolutionary advancements that AIoT can bring to myriad sectors. Thoughtful and strategic planning, combined with innovative design, forms the cornerstone upon which successful AIoT solutions are built, paving the way for a smarter, interconnected future.

Deployment Strategies

Successfully implementing AIoT solutions requires a well-thought-out deployment strategy that takes into account the unique characteristics and demands of both AI and IoT systems. The deployment phase is crucial, as it translates the conceptual and design stages of AIoT projects into functional, real-world applications. Without careful attention to deployment strategies, even the most promising AIoT solutions might fail to deliver their intended benefits.

The first step in any deployment strategy is to establish clear objectives. This involves identifying the specific problems that the AIoT solution aims to solve and determining measurable goals. Whether improving operational efficiency, enhancing user experience, or providing predictive maintenance, these goals guide the deployment process and help in evaluating success post-implementation.

For a deployment strategy to be effective, it must address scalability. AIoT solutions often start with a small-scale pilot project, but the ultimate aim is to extend them across larger systems or geographic areas. Scaling up involves not only technical considerations but also logistical planning, such as ensuring that there is adequate infrastructure to support the expanded system. Scalability also includes the flexibility to adapt to new data sources and evolving user requirements without significant overhauls.

Interoperability is another critical element in deploying AIoT solutions. Given the variety of devices, platforms, and communication protocols involved in IoT, ensuring that these components can work together seamlessly is paramount. Using standard protocols and open-source platforms where possible can ease the integration process, making it simpler to add new devices or systems to the network without extensive reconfiguration.

Security is non-negotiable when it comes to deploying AIoT systems. From the outset, the deployment strategy must include robust cybersecurity measures to protect against threats such as data breaches, unauthorized access, and cyberattacks. This involves implementing encryption, secure authentication protocols, and regular security audits. Given that AI systems often handle sensitive data, ensuring data integrity and confidentiality throughout the data lifecycle is essential.

Monitoring and maintenance are crucial aspects of any deployment strategy. Once an AIoT solution is live, continuous monitoring ensures that the system operates as expected and identifies any issues before they become critical. This can include real-time performance monitoring, predictive maintenance alerts, and periodic system audits. Regular updates and patches are also necessary to keep the system current and secure against emerging threats.

Cost considerations should not be overlooked. Budget constraints can influence the choice of technologies, the scale of initial implementation, and long-term sustainability of the AIoT solution. A thorough cost-benefit analysis helps in making informed decisions about the most cost-effective approaches without compromising on quality and functionality. Often, investing in scalable and modular technologies can offer long-term savings.

User training and stakeholder engagement are often underemphasized but play a vital role in the deployment phase. Ensuring that end-users and stakeholders are well-informed about the functionalities and benefits of the AIoT system can facilitate smoother adoption and greater enthusiasm for the new technology. Training programs, workshops, and comprehensive user manuals can significantly reduce resistance and enhance user experience.

Moreover, adopting an iterative deployment approach can provide significant advantages. Iterative methods allow for incremental improvements and continuous feedback, making it easier to adapt and refine the solution based on real-world performance and user feedback. This approach is particularly useful for AI components, which typically benefit from ongoing learning and optimization.

Risk management is another critical consideration. Anticipating potential risks, such as technological failures, cybersecurity threats, and regulatory challenges, helps in developing contingency plans. These plans ensure that the system can quickly recover from disruptions and continue to meet its objectives. Risk assessments should be a continuous process, even after the deployment, to adapt to new challenges and minimize adverse impacts.

Finally, it's essential to establish key performance indicators (KPIs) to measure the success of the AIoT deployment. KPIs should align with the initial objectives and provide quantifiable metrics that can be regularly reviewed to assess performance. Metrics might include system uptime, data accuracy, user satisfaction, and ROI. By continuously monitoring these indicators, organizations can make data-driven decisions to optimize and scale their AIoT solutions effectively.

In conclusion, the deployment of AIoT solutions is a multifaceted process requiring careful planning, robust security measures, continuous monitoring, and an iterative approach. A successful deployment strategy not only brings the AIoT solution to life but also ensures that it remains functional, secure, and scalable in the long term. By focusing on these key elements, organizations can harness the full potential of AIoT to revolutionize their operations and deliver transformative benefits across various domains.

Case Implementation

Implementing AIoT solutions requires a well-thought-out approach that bridges the gap between conceptual planning and practical execution. This process is the linchpin in transforming theoretical designs into operational systems that deliver tangible results. Let's delve into how various industries have successfully navigated the challenging waters of AIoT implementation.

One exemplary case is the application of AIoT in agriculture, specifically in precision farming. Precision farming uses sensors and IoT devices to collect data on soil conditions, crop health, and weather patterns. Artificial Intelligence analyzes this data to provide actionable insights for farmers. For example, an agritech company implemented an AIoT solution that fused drone technology with AI algorithms to monitor crop health in real-time. They achieved this by deploying drones equipped with high-resolution cameras and sensors over large farming areas. These drones collected vast amounts of data, which were then processed by machine learning algorithms to detect signs of disease, pest infestation, and nutrient deficiencies in crops.

The real challenge was not just in deploying the technology but in making it user-friendly for farmers who might not be tech-savvy. Therefore, the data insights were presented through a simple mobile app interface, allowing farmers to take immediate action based on the recommendations. This implementation reduced pesticide use by 30% and increased crop yield by 15%, proving the efficacy of their AIoT system.

In the retail industry, AIoT solutions are transforming the customer experience and inventory management processes. A large retail chain successfully implemented AIoT to streamline its supply chain operations. They used IoT sensors in warehouses to monitor the stock levels in real-time and AI algorithms to predict consumer demand accurately. These predictions were based on factors such as seasonality, current trends, and historical sales data. To execute this, the retailer integrated IoT sensors capable of collecting real-time data on inventory status, shelf life, and product movement.

The data collected was then fed into AI systems that used machine learning models to forecast demand with high accuracy. This approach enabled the chain to optimize stock levels, reducing both overstock and stockouts. Additionally, by implementing autonomous robots for inventory checks, they reduced manual labor costs and increased efficiency. This combination of IoT data and AI-driven analytics created a robust, responsive supply chain capable of adapting to market fluctuations in near real-time.

Healthcare is another arena where AIoT implementation is making groundbreaking strides. Consider the case of remote patient monitoring. A healthcare provider integrated IoT wearable devices that tracked vital signs such as heart rate, blood pressure, and blood oxygen levels. These devices were connected to an AI platform that continuously analyzed patient data, detecting anomalies and predicting potential health issues before they became critical. The deployment strategy involved ensuring that the wearable devices were comfortable, reliable, and capable of long-term use without frequent recharging.

The AI system was designed to alert healthcare professionals and caregivers through a secure mobile app, providing them with real-time insights into the patient's health status. This allowed for timely interventions and personalized care plans, substantially reducing hospital readmissions and improving patient outcomes. Moreover, the data collected offered a long-term view of the patient's health, enabling doctors to make more informed decisions.

In the realm of intelligent cities, AIoT solutions are being implemented to enhance urban living. A noteworthy example is a city's effort to optimize its public transportation system. The project involved the installation of IoT sensors across various modes of transport—buses, trains, and trams—to collect data on passenger counts, travel times, and congestion levels. This data was integrated into an AI system designed to analyze current traffic conditions, predict future patterns, and optimize routes and schedules accordingly.

Key to the success of this implementation was the seamless integration of AI analytics with existing city infrastructure. The city utilized a combination of edge computing and cloud services to process large volumes of data efficiently. Transit authorities and commuters accessed the AI-generated recommendations through digital signages and mobile apps, providing real-time updates and alternative routes. This implementation not only improved traffic flow but also reduced commuting time and decreased the carbon footprint by optimizing fuel use.

Manufacturing companies have also harnessed AIoT for predictive maintenance and operational efficiency. A leading automotive manufacturer implemented an AIoT system to monitor the health of its assembly line machinery. IoT sensors were installed on critical components to measure variables like temperature, vibration, and wear. The AI algorithms analyzed this data to predict equipment failures before they occurred, enabling proactive maintenance scheduling.

This case emphasized the importance of data quality and integrity. Ensuring the IoT sensors' accuracy and the reliability of the data pipeline was crucial for the system's effectiveness. The company created a rigorous data validation process to filter out noise and ensure high-quality inputs for their machine learning models. Consequently, they saw a reduction in unscheduled downtimes by 40%, significantly improving their production throughput and operational efficiency.

Another prime example is AIoT applications in energy management, particularly in smart grids. An energy provider implemented smart meters and IoT sensors across its grid to feed real-time data into an AI system for demand forecasting and resource allocation. The AI analyzed consumption patterns, weather data, and grid performance metrics to optimize energy distribution dynamically.

In implementing this solution, the provider faced challenges related to data latency and integration with legacy systems. They addressed these by deploying edge computing solutions to process data closer to its source, thereby reducing latency. The success of this project hinged on robust data integration protocols that ensured seamless communication between the new AIoT system and existing grid management systems. This resulted in improved grid reliability, reduced energy wastage, and substantial cost savings for both the provider and its customers.

These case implementations highlight that the journey from planning to execution involves multiple layers of complexity, each requiring careful consideration and strategic problem-solving. It's not just about deploying sensors or analytics platforms but about creating an ecosystem where these components can work in harmony to deliver value. Understanding the unique challenges and requirements of each industry is crucial to designing AIoT solutions that are not only innovative but also practical and scalable.

While the benefits of AIoT are manifold, it's essential to approach implementation with a mindset geared towards continuous improvement. Each deployment should be viewed as a learning opportunity, with frequent iterations based on real-world feedback. This iterative process ensures that the solutions evolve to meet changing needs and technological advancements, ultimately leading to more robust and resilient systems.

In summary, successful AIoT implementation requires a nuanced understanding of the specific industry challenges, meticulous planning, and a willingness to adapt. As more industries adopt these technologies, the insights gleaned from these implementations will pave the way for even more transformative applications, shaping a future where AI and IoT are integral to our daily lives and industrial processes.

CHAPTER 24:
COLLABORATION AND ECOSYSTEM

Integrating AI and IoT seamlessly into our world requires more than just cutting-edge technology; it demands robust collaboration and an ecosystem that fosters innovation and mutual growth. Partnerships and alliances among tech giants, startups, and industry leaders accelerate advancements by combining unique strengths and resources. Developer communities thrive as open source initiatives encourage transparency and shared knowledge, breaking down barriers to entry and spurring creative solutions. In this interconnected landscape, each player from individual developers to multinational corporations, contributes to a thriving environment where best practices and breakthrough technologies emerge. The synergy between collaboration and ecosystem not only propels AIoT forward but also ensures these advancements remain inclusive, scalable, and impactful across diverse sectors.

Partnerships and Alliances

At the heart of driving innovation in AI and IoT lies the strength and breadth of partnerships and alliances. These collaborations serve as a critical mechanism for both research and commercialization, fostering ecosystems that spur growth, knowledge sharing, and technological advancements. As AI and IoT continue to permeate various sectors, the significance of strategic alliances cannot be overstated.

Charlie Morgan

The technological landscape is complex, and no single entity can master every aspect of AI or IoT. It is through partnerships that companies can leverage each other's unique capabilities and expertise. Take, for example, the synergy between hardware manufacturers and software developers. The hardware provides the essential foundation for IoT devices, while sophisticated AI algorithms transform raw data into actionable insights. By working together, companies can introduce transformative solutions that neither could achieve alone.

Collaborations between academia and industry also play a pivotal role. Universities are hotbeds of innovation, where cutting-edge research into AI and IoT is conducted. By aligning with academic institutions, businesses gain access to pioneering research, fresh talent, and novel technologies. In return, academic institutions benefit from real-world data, industry insights, and funding opportunities for further research. This symbiotic relationship accelerates the pace of innovation and ensures that theoretical advancements quickly find practical applications.

Another vital aspect of partnerships is the joint ventures between established companies and startups. Startups bring agility, creativity, and novel approaches to problem-solving. Established companies contribute with their vast resources, market reach, and industry experience. When these two entities collaborate, the result is often groundbreaking. For instance, an established IoT service provider might pair with a startup specializing in AI to create a robust, AI-enabled IoT platform, blending the startup's cutting-edge technology with the veteran company's market expertise.

Similarly, cross-industry partnerships are becoming increasingly common. Traditional industry boundaries are dissolving as companies from diverse sectors join forces to tackle complex challenges. An automaker, an AI firm, and a telecommunications company might collaborate on developing connected, autonomous vehicles. Each partner contributes their specific domain knowledge, transforming the way we think about transportation.

In addition to formal partnerships, informal alliances are equally impactful. Participation in industry consortia and standards bodies allows companies to collectively shape the direction of AI and IoT development. These bodies are crucial for establishing industry standards, which promote interoperability and ensure seamless integration of devices and systems. The collective voice of these alliances is often stronger than individual companies, enabling more rapid and widespread adoption of technologies.

Moreover, public-private partnerships (PPPs) have emerged as powerful tools for advancing AI and IoT initiatives that serve the public good. Governments partner with private companies to deploy smart city projects, enhance public safety, and improve healthcare delivery. These collaborations bring together the innovative prowess of the private sector with the regulatory and financial support of the public sector, resulting in substantial societal benefits.

Partnerships also play a critical role in addressing ethical and privacy concerns. By working together, companies can develop best practices and guidelines that ensure ethical AI deployment and robust IoT security. Consortia focused on ethical AI can pool resources to research and address common challenges, such as bias in algorithms or privacy in IoT systems. Through collaboration, these groups develop frameworks that each member can adopt, leading to industry-wide improvements.

Global partnerships are equally influential. As AI and IoT are not confined by geographical boundaries, international collaborations offer a pathway to a more cohesive global technology landscape. Companies from different parts of the world bring diverse perspectives and expertise, enriching the innovation process. Global alliances facilitate knowledge transfer, standardization, and the deployment of technologies in diverse contexts, broadening the impact of AI and IoT innovations.

For instance, the advent of 5G technology calls for robust international cooperation. Telecommunications giants across continents are working together to ensure that AI and IoT devices can seamlessly connect to high-speed, low-latency networks. This collaborative effort is vital for realizing the full potential of emerging technologies, from autonomous vehicles to smart grids.

In the same vein, multinational corporations often spearhead initiatives to create global solutions addressing universal challenges, such as climate change. Collaborations among energy companies, AI firms, and international organizations can lead to the development of smart grids and AI-driven renewable energy solutions. These solutions not only promote sustainability but also contribute to achieving global environmental goals.

Another significant dimension of partnerships is in fostering developer communities. Developer ecosystems serve as the breeding ground for innovation, where developers from various organizations collaborate, share insights, and build upon each other's work. By supporting developer communities, companies ensure a continuous exchange of ideas, leading to innovations that can be quickly adopted and scaled.

Open source initiatives are another exciting area where partnerships flourish. By contributing to and drawing from open source projects, companies can leverage a vast pool of talent and accelerate their development processes. Open source communities bring together developers from all over the world to build, improve, and maintain software that drives AI and IoT advancements. This collaborative approach democratizes technology, making cutting-edge solutions accessible to a broader audience.

Furthermore, partnerships with regulatory bodies are essential to navigate the complex legal landscape surrounding AI and IoT. Companies partner with regulators to stay compliant with evolving laws and contribute to the development of new regulations that facilitate innovation while protecting public interest. These partnerships ensure that the regulatory environment evolves in ways that support technological growth without compromising ethical standards and public trust.

To sum up, partnerships and alliances are the bedrock upon which the future of AI and IoT will be built. They span across different types of organizations, industries, and even countries, each bringing a unique set of strengths to the table. Through collaborative efforts, these partnerships not only drive innovation but also ensure that technological advancements are deployed in ways that are ethical, sustainable, and beneficial to society as a whole.

As AI and IoT continue to evolve, the role of partnerships and alliances will only become more critical. The fusion of diverse expertise and resources will pave the way for innovations we can only begin to imagine, shaping a future where technology works seamlessly to enhance every aspect of our lives.

Developer Communities

Developer communities are the lifeblood of innovation in the rapidly unfolding era of AI and IoT. These communities, comprising both seasoned professionals and enthusiastic novices, create a vibrant ecosystem where collaboration thrives and burgeoning ideas can take root. Fundamentally, these groups are pivotal for knowledge exchange, skill development, and the collective advancement of technology. By sharing their diverse experiences and innovative solutions, they push the boundaries of what AI and IoT can achieve, collectively solving complex problems that single entities would struggle to tackle alone.

Within these communities, the collaborative spirit reigns supreme. Whether through open-source projects, hackathons, or online forums, developers find a platform to voice their ideas, seek feedback, and refine their solutions. GitHub, for instance, stands as a monumental hub for open-source collaboration. Repositories teeming with code samples, documentation, and walkthroughs enable developers to contribute to and draw from a pool of shared knowledge. This mutual exchange accelerates innovation, as individuals can build upon each other's work without starting from scratch.

The beauty of developer communities transcends geographical boundaries. Thanks to digital platforms, developers from different parts of the world can work together in real-time or asynchronously. Tools like Slack, Discord, and developer-centric social media channels such as Stack Overflow or Reddit contribute to an ecosystem where questions are answered promptly, and new partnerships are forged. These platforms offer not just solutions but also a sense of camaraderie among like-minded individuals, driving a collective mission to harness the full capabilities of AI and IoT.

Certainly, hackathons embody the creative dynamism and collaborative ethos of developer communities. These events, which can be in-person or virtual, bring together developers, designers, and entrepreneurs to collaborate intensively over short periods, solving predefined challenges or innovating within specified themes. Hackathons often focus on utilizing AI and IoT to address real-world issues, from urban planning and healthcare to environmental sustainability. Participants benefit not only from fresh perspectives but also from mentoring by industry experts, accelerating both their learning and their projects.

Professional networking and mentorship are other crucial aspects bolstered within developer communities. Seasoned developers and industry veterans often participate in forums, webinars, and local meetups, offering their wealth of experience to guide newcomers. These interactions can demystify complex concepts and provide career advice and leadership insights, enriching the overall skill set of the community. The mentoring dynamic ensures a continuous influx of capable and inspired developers ready to push the envelope in AI and IoT.

Another significant driver within developer communities is the array of specialized sub-communities or groups focusing on niche aspects of AI or IoT. For instance, there may be clusters of developers dedicated specifically to AI ethics, IoT security protocols, or AI-driven healthcare solutions. These specialized groups dive deep into their areas of interest, creating more focused discussions and highly tailored resources. The granularity of these communities allows for an in-depth exploration of specific issues, fostering an environment where innovation can flourish unfettered by broader, less focused dialogue.

Additionally, academic institutions play a pivotal role in fostering robust developer communities. Universities often collaborate with tech companies to host coding bootcamps, research symposiums, and independent study programs aimed at enhancing student engagement in AI and IoT. Students are encouraged to join or form developer clubs, participate in hackathons, and contribute to open-source projects. These initiatives not only prepare students for the professional world but also ensure a constant influx of fresh ideas and perspectives into the broader developer ecosystem.

The industry support for these communities cannot be overstated. Major tech companies like Google, Microsoft, and IBM have robust developer support programs, often providing resources, APIs, and technical support to grassroots developer initiatives. These corporations recognize the value developer communities bring in terms of early adoption, product feedback, and brand advocacy. By investing in these communities, companies can foster goodwill, encourage innovation, and stay attuned to the cutting edge of technology development.

In this collaborative universe, the feedback loop is of utmost importance. Developers contribute innovations, but they also receive invaluable feedback from their peers. This iterative process ensures that solutions are continually refined and improved. Open-source platforms particularly benefit from this model, as numerous contributors review, test, and enhance the code, leading to robust and flexible solutions that can be quickly adapted and scaled.

Equally important is the role of conferences and workshops. Events like the AIoT Dev Summit or the International Conference on IoT and AI offer unique opportunities for developers to present their work, attend expert-led sessions, and network with peers and industry leaders. These gatherings are fertile ground for inspiration, offering a blend of theoretical knowledge and practical insights. Attending these events can ignite new ideas, hone existing projects, and even open doors to collaborative ventures that carry forward long after the conference ends.

Furthermore, the contribution of developer communities to establishing and maintaining best practices can't be ignored. Through shared experiences and collective wisdom, they help define the standards and protocols that ensure interoperability, security, and functionality in AI and IoT solutions. These best practices are often documented in shared repositories, FAQs, and tutorial videos, forming an ever-evolving guidebook for both novice and expert developers. This shared knowledge base ensures that new solutions adhere to established guidelines, improving both the reliability and the safety of implemented technologies.

In the ongoing dialogue about ethics and data privacy, developer communities serve as vital stakeholders. They provide a platform for discussing and debating the ethical implications of AI and IoT developments. This communal brainstorming is crucial for establishing norms and guiding principles that developers worldwide can adopt. The cross-pollination of ideas within these forums often leads to more holistic, ethically sound approaches to technology deployment.

The democratization of technology is another profound impact of developer communities. By making resources, tools, and knowledge freely accessible, these communities break down barriers to entry. This democratization is pivotal in ensuring diverse participation, fostering inclusivity, and tapping into a broader range of insights and innovations. It allows individuals from varying backgrounds—whether educational, cultural, or economic—to contribute to and benefit from advancements in AI and IoT.

Developer communities also offer a launchpad for entrepreneurial ventures. Often, the innovative ideas generated within these groups lead to the formation of startups. Developers find co-founders, investors, or early adopters within these communities, turning collaborative projects into viable business models. This entrepreneurial spirit fuels economic growth and spurs further innovation, creating a virtuous cycle that benefits the entire ecosystem. Startups emerging from these collaborative spaces are often celebrated for their agility and innovative edge, often bringing novel solutions to market at an accelerated pace.

In summation, developer communities stand at the heart of the collaboration and ecosystem of AI and IoT. They embody the spirit of collective innovation, bridging gaps across geographies, industries, and expertise levels. By fostering a dynamic and inclusive environment for sharing knowledge, refining ideas, and launching new initiatives, these communities significantly contribute to the rapid evolution and responsible application of AI and IoT technologies. As we look to the future, the strength and vibrancy of these communities will undoubtedly continue to play a critical role in shaping the landscape of technological advancement.

Charlie Morgan

Open Source Initiatives

The essence of open-source initiatives lies in a community-driven approach to technology development. These initiatives thrive on the principles of transparency, collaboration, and shared growth. By making software and hardware designs publicly accessible, they democratize innovation, allowing anyone to contribute to and benefit from technological advancements. This paradigm is particularly transformative in the context of AI and IoT integration, where the complexity and rapid pace of development can otherwise create barriers to entry.

One of the most significant advantages of open-source projects is their ability to foster global collaboration. Developers from around the world can work together to enhance functionalities, fix bugs, and introduce new features more rapidly than any single organization could manage. This collective effort often results in more robust, versatile, and reliable solutions. For instance, TensorFlow, an open-source machine learning framework developed by Google, has seen contributions from thousands of developers, leading to its widespread adoption and continual improvement.

In the realm of IoT, platforms like Node-RED and OpenHAB exemplify the power of open-source communities. Node-RED, developed by IBM, is a flow-based development tool for visual programming, enabling users to wire together devices, APIs, and online services. Its open-source nature allows for the addition of a vast array of nodes, extending its capabilities and fostering an ecosystem where innovation flourishes. OpenHAB, on the other hand, provides an integrative platform for home automation. It supports numerous IoT devices and systems, promoting a level of interoperability that proprietary solutions often fail to achieve.

Such initiatives don't just benefit developers; they have a profound impact on industries and end-users as well. Open-source solutions often lead to cost reductions, as businesses can implement and customize them without the hefty licensing fees associated with proprietary software. This is particularly beneficial for startups and small enterprises that might otherwise struggle to afford the latest technological advancements.

The accessibility of open-source software can also accelerate research and development within academic and scientific communities. Researchers can build upon existing frameworks without reinventing the wheel, allowing them to focus on innovation rather than basic infrastructure. Kaggle, a free platform for predictive modeling and analytics competitions, often provides access to open-source datasets and tools, enabling data scientists and AI researchers to test novel hypotheses and algorithms in a collaborative environment.

However, the value of open-source initiatives extends beyond software. Open hardware projects, such as the Arduino microcontroller and the Raspberry Pi computer, have revolutionized the way we prototype and develop IoT devices. These platforms empower hobbyists, educators, and professionals alike to experiment, learn, and bring their ideas to life without prohibitive costs. The community-centric nature of these projects often leads to extensive documentation, tutorials, and forums where users can seek assistance and share their experiences.

One notable example of the impact of open hardware is the Internet of Things. By using widely accessible platforms like Arduino and Raspberry Pi, developers can create custom IoT solutions tailored to their specific needs. This flexibility has catalyzed innovation in diverse fields, from smart agriculture to advanced healthcare solutions. In agriculture, for instance, open-source IoT devices can monitor soil moisture and weather conditions, enabling precise irrigation and optimizing crop yields.

It's not just the tech giants and developers who are benefiting from open-source initiatives. Governments and public institutions are also increasingly adopting open-source solutions to enhance transparency, reduce costs, and foster innovation. The adoption of open data policies enables the creation of smart city solutions that leverage AI and IoT for urban planning, traffic management, and public safety. By making data publicly accessible, cities like Barcelona and Amsterdam have empowered citizens and developers to create innovative solutions that improve the quality of life.

Moreover, the open-source ethos aligns perfectly with the drive for ethical AI practices. The transparency inherent in open-source projects facilitates peer review and auditing, which are crucial for ensuring the fairness, accountability, and transparency of AI systems. Organizations can scrutinize algorithms to mitigate biases and enhance trust in AI-driven decisions. This open scrutiny is essential in industries like healthcare, where AI algorithms assist in diagnostics and treatment planning, and any biases or errors could have dire consequences.

While the benefits of open-source initiatives are manifold, they are not without challenges. One primary concern is the sustainability of open-source projects, which often rely on the contributions of volunteers and the support of a few core maintainers. Ensuring that critical projects receive the necessary funding and resources is essential to their longevity and success. Some projects address this by seeking sponsorships, donations, or adopting dual licensing models where commercial use requires a paid license.

Additionally, the open nature of these projects can sometimes lead to fragmentation, where multiple versions or forks of the same software exist, each with slight variations. While this can foster innovation, it can also complicate collaboration and integration efforts. Establishing clear guidelines and governance structures can help mitigate these issues, ensuring that the main project remains cohesive even as it evolves.

The role of large corporations in the open-source ecosystem is also evolving. Many tech giants, such as Microsoft and IBM, contribute significantly to open-source projects, recognizing their strategic importance. While this collaboration between corporate and community interests can drive substantial advancements, it also raises questions about the influence and control these corporations may exert over the direction of open-source projects. Maintaining a balance where community interests are not overshadowed by corporate agendas is crucial to preserving the integrity of the open-source ethos.

Looking to the future, the integration of AI and IoT within the framework of open-source initiatives holds the potential to revolutionize technology further. Innovations such as federated learning, where AI models are trained across decentralized devices without sharing raw data, could benefit immensely from open-source collaborations. These initiatives can develop standards and protocols that facilitate secure and efficient implementation, addressing privacy concerns while harnessing the power of distributed data.

Moreover, as the open-source community continues to grow, we can anticipate more interdisciplinary collaborations that bring together experts from AI, IoT, cybersecurity, and ethics to address complex challenges. This cross-pollination of ideas will not only foster technological innovation but also ensure that the solutions developed are holistic and considerate of societal impacts.

Ultimately, the spirit of open-source initiatives embodies a collective endeavor to push the boundaries of what is technologically possible while ensuring that the benefits of innovation are accessible to all. In the age of AI and IoT, this collaborative and inclusive approach will be instrumental in shaping a future where technology enhances and enriches lives across the globe. As we advance, continuing to nurture and support open-source initiatives will be key to unlocking the full potential of AIoT and ensuring its positive impact on society.

CHAPTER 25:
ENHANCING USER EXPERIENCE
WITH AIoT

As we navigate the future shaped by the integration of Artificial Intelligence and the Internet of Things, enhancing user experience becomes paramount. By incorporating user-centered design principles, AIoT-driven solutions can intuitively anticipate user needs and preferences. From sophisticated AI in customer support, which provides real-time assistance and personalized interactions, to improved usability and accessibility features that ensure inclusive tech adoption, AIoT is revolutionizing how we interact with the world around us. Not only does it streamline complex processes, but it also creates seamless, efficient, and genuinely enriching experiences for users across diverse contexts and industries. The impact on daily life and industry is profound, driving forward a more intelligent, responsive, and user-friendly technological landscape.

User-Centered Design

In the ever-evolving landscape of technology, focusing on user-centered design (UCD) becomes crucial when integrating Artificial Intelligence (AI) with the Internet of Things (IoT). The essence of UCD lies in crafting products and services that are tailored to meet the specific needs and expectations of end-users. As AIoT solutions permeate various aspects of our lives—from smart homes to industrial settings—the role of UCD becomes even more compelling.

Why is user-centered design so important in AIoT? Well, it's simple. Users interact with technology on a daily basis; their experience determines whether these interactions are seamless or fraught with frustration. A well-designed AIoT system should feel natural and intuitive. Any friction in user experience can result in lost productivity, dissatisfaction, and ultimately, abandonment of the technology altogether.

However, achieving a user-centered design for AIoT solutions isn't straightforward. It requires a deep understanding of the users—what they need, what they value, their capabilities, and their limitations. This insight is often gathered through user research methods such as surveys, interviews, and usability tests. By genuinely understanding users, designers can create systems that not only perform well but are also pleasant and easy to use.

Additionally, UCD in AIoT is not a one-time activity; it's an iterative process. Initial designs are tested with users, insights are gathered, and refinements are made. This cycle continues until the product reaches a point of optimal usability. Moreover, even after deployment, continuous feedback and updates ensure the system evolves according to user requirements and technological advancements.

An interesting aspect of integrating AI into IoT devices is the potential for adaptive learning. AI algorithms can monitor how users interact with the system and then adapt dynamically to enhance the user experience. For example, a smart thermostat could learn a user's temperature preferences over time and adjust settings accordingly, making the home environment more comfortable without manual adjustments.

One fundamental principle of UCD is accessibility. Not all users have the same abilities, and technology should be designed to be as inclusive as possible. AIoT solutions should be accessible to everyone, including those with disabilities. Voice-activated commands, adjustable font sizes, and alternative text for images are just a few ways to make technology more accessible. AI can further contribute by providing personalized accessibility features based on user behavior and preferences.

Consider the smart home ecosystem. Users require more than just functional devices; they need systems that are easy to setup, manage, and interact with daily. UCD for smart homes involves simplifying complex configurations and ensuring devices communicate effortlessly with one another. In this scenario, AI plays a critical role by automating routine tasks—like turning off lights when no one is in the room—thereby enhancing user comfort and convenience.

Another area where UCD proves pivotal is in industrial settings. Here, the stakes are often higher due to the complexity and scale of operations. User-centered design focuses on creating intuitive interfaces for operators, reducing the risk of errors, and improving efficiency. Predictive maintenance systems, powered by AI, alert operators before machinery malfunctions, allowing for timely interventions and minimizing downtime.

The benefits of UCD in AIoT extend to customer service as well. AI-driven chatbots and virtual assistants exemplify how seamless interaction can enhance user satisfaction. When these systems are designed with the user in mind—considering factors like natural language processing, quick response times, and problem-solving capabilities—they become invaluable tools for customer engagement and support.

Furthermore, scalability and personalization are essential when discussing user-centered design in AIoT. As users' needs grow and diversify, the systems must scale accordingly and provide personalized experiences. For instance, in a healthcare setting, wearable devices that monitor vital signs should cater to individual healthcare plans, adapting recommendations and alerts based on personal health data analyzed by AI.

User feedback loops are critical in this context. A robust feedback mechanism ensures that users can continuously share their experiences and suggestions. This feedback is invaluable for making iterative improvements and ensuring that AIoT solutions evolve in line with user needs. Additionally, integrating real-time analytics allows for the dynamic adaptation of functionalities, making the system more responsive and user-centric.

Innovation in user-centered design also often comes from unexpected places. Leverage cross-disciplinary insights—from psychology to sociology—to create more empathetic and effective designs. Involving diverse stakeholders in the design process brings in multiple perspectives, enriching the final AIoT system with features and functionalities that genuinely resonate with users.

In conclusion, user-centered design is not just an add-on; it's a fundamental approach that aligns technology with human needs and expectations. By placing users at the heart of AIoT development, we can create smarter, more efficient systems that enhance everyday life and redefine industrial processes. Beyond functionality, UCD ensures that these advances are also broadly accessible, ultimately leading to a more inclusive technological future.

AI in Customer Support

The landscape of customer support has transformed dramatically with the incorporation of Artificial Intelligence (AI). Traditional methods often involved long wait times, repetitive queries, and human errors, creating a less than ideal user experience. Now, leveraging AI within the realm of the Internet of Things (IoT) has turned this scenario on its head, enhancing user satisfaction, streamlining operations, and even predicting and resolving issues before they occur.

One of the most impactful advancements is the deployment of AI-powered chatbots. These digital assistants are capable of handling an array of customer inquiries, from the simple to the complex. Unlike their human counterparts, AI chatbots can operate 24/7 and manage multiple interactions simultaneously. This not only reduces waiting times but also ensures that customer queries are addressed in real time. It is not just about handling volume; the quality of interaction is equally important. Modern chatbots employ Natural Language Processing (NLP) to understand and respond to queries in a conversational manner, making the interaction feel personal and more intuitive.

Incorporating IoT data into these AI systems further refines customer support. Imagine a scenario where a smart home device, like a thermostat, detects an unusual activity or malfunction. Through AI and IoT integration, the device can self-diagnose the issue and communicate it to a customer support system. The system, equipped with AI algorithms, can then guide the customer through troubleshooting steps or automatically schedule a service appointment if needed. This doesn't just save time; it also preempts potential dissatisfaction by addressing issues proactively.

An essential aspect of AI in customer support is personalization. AI systems analyze data from various IoT devices to comprehend user preferences and behaviors. This allows for a tailored customer experience. For example, if a user frequently queries about specific functionalities of a device, the AI system learns and adapts, offering more relevant information and suggestions in future interactions. Personalization extends beyond troubleshooting; it encompasses advice, recommendations, and upsell opportunities tailored to individual needs, thus enhancing the overall user experience.

Moreover, AI in customer support is invaluable for monitoring and analyzing customer sentiments. Advanced algorithms can sift through vast amounts of interaction data to gauge customer satisfaction and identify pain points. This enables businesses to act swiftly in addressing issues and refining their products or services. When coupled with IoT data, the insights gained are deeper and more actionable. For instance, if a significant number of users exhibit frustration with a particular feature of a connected device, the company can prioritize updates or provide focused training materials to enhance usability.

The predictive capabilities of AI bring a transformative edge to customer support. Leveraging machine learning algorithms, AI systems can forecast potential issues before they occur. This is particularly beneficial in industries where downtime can be critical, such as in manufacturing or healthcare. By analyzing patterns from IoT devices, AI can predict equipment failures or maintenance needs, enabling preemptive action. This not only minimizes disruptions but also extends the lifespan of the equipment and improves safety standards.

Another critical advantage of integrating AI in customer support is the scalability it offers. As a business grows, the demand for customer support typically escalates. Scaling human-operated support teams proportionally can be costly and impractical. AI systems, on the other hand, are inherently scalable. They can handle increasing volumes of interactions without the need for corresponding increases in human resources, thus providing a cost-effective solution that ensures consistent support quality.

On the backend, AI also assists human agents by providing them with relevant information and suggestions during customer interactions. This hybrid approach leverages the best of both worlds—the efficiency and accuracy of AI and the empathy and understanding of human touch. AI can quickly retrieve and present relevant data from IoT sensors or historical interaction logs, enabling agents to resolve issues more effectively and efficiently. This support allows human agents to focus on more complex issues that require nuanced understanding and personalized attention.

It's also worth noting the role of AI in continuous learning and improvement within customer support systems. AI models are designed to learn from each interaction, continuously improving their accuracy and relevance over time. This means that the more the system is used, the better it becomes at understanding and addressing customer needs. Feedback loops where users rate their support experience further refine these models, ensuring that the AI not only reacts but evolves.

Furthermore, AI in customer support enhances data security and privacy. With increasing concerns over data breaches and unauthorized access, AI solutions are equipped with advanced security protocols to protect sensitive customer information. They can also monitor for unusual patterns that might indicate security threats, allowing for quick intervention. This is particularly critical when dealing with IoT devices that collect vast amounts of personal and operational data.

The integration of AI in customer support is more than just a technological enhancement—it's a paradigm shift that redefines how businesses interact with their customers. It's about creating a seamless, proactive, and deeply personalized user experience that anticipates needs and resolves issues with minimal friction. More importantly, it signifies a move towards a more sustainable customer service model where efficiency is balanced with empathy, leverage is maximized through continuous learning, and security is integral.

In conclusion, the seamless integration of AI and IoT in customer support is already a game-changer and promises to be even more transformative in the future. By harnessing the power of real-time data, predictive analytics, and advanced machine learning, businesses can not only meet their current customer service demands but also anticipate and adapt to future needs. As AI continues to evolve, the potential to further enhance customer support and, consequently, user experience, expands exponentially, charting a course toward a more intelligent, responsive, and user-centered technological ecosystem.

Usability and Accessibility

In our increasingly digital world, designing user experiences that prioritize usability and accessibility is more important than ever. When we talk about enhancing user experience with AIoT (the integration of Artificial Intelligence and the Internet of Things), we must focus on creating systems that are not just functional but also intuitively easy to use for everyone, irrespective of their abilities.

It all starts with a user-centered design approach. This means understanding the needs and challenges faced by the user and then creating solutions that align with those insights. AIoT allows for significant personalization, adapting devices and interfaces to meet the unique preferences of each individual user. For instance, a smart home system can learn a resident's daily habits and adjust lighting, temperature, and security settings automatically.

Still, personalization alone is not enough. We must think about accessibility. Accessible design ensures that people with disabilities can use AIoT devices and services with ease. A key aspect of this is voice recognition technology, which allows users to control devices without needing to physically interact with them. This is particularly beneficial for individuals with mobility impairments. Additionally, visual and auditory aids, such as screen readers and closed captioning, can make digital content more accessible for users with vision and hearing impairments.

To achieve high usability, the interfaces of AIoT devices must be intuitive. The complexity of technology should be hidden from the user; they should interact with the system seamlessly and effortlessly. User testing is crucial here. Iterative testing with diverse user groups helps identify usability issues and areas for improvement. Furthermore, contextual inquiry—where designers observe users in their natural environment—can provide invaluable insights.

AIoT can significantly improve accessibility by leveraging AI to provide adaptive interfaces. For example, an AI assistant could adjust text sizes, color contrasts, and audio outputs based on the user's preferences or needs detected over time. This adaptability ensures that the system remains usable as the user's requirements evolve.

Usability isn't just about ease of use; it's also about making sure the user feels confident and secure while interacting with technology. Transparency in how an AIoT system collects and uses data can build user trust. Clear, simple language explaining the actions and purposes of AI functions can demystify technology and reduce hesitation or uncertainty.

Moreover, the importance of robust error handling cannot be overstated. Users should know what went wrong and how they can rectify it. Educational prompts and error messages that guide users through troubleshooting can enhance the overall user experience. This not only improves usability but also helps in building a user's confidence in using AIoT devices.

Beyond usability and accessibility, ethical considerations are paramount in the development of AIoT systems. This involves designing systems that are not biased and are inclusive of all demographics. Inclusive design not only caters to the needs of people with disabilities but also accommodates the broadest range of human diversity. This might involve considering various ethnic, cultural, and linguistic backgrounds in the design process.

Imagine a smart healthcare system that provides personalized patient care but is accessible to elderly patients who may not be tech-savvy. The interface could feature larger buttons, voice instructions, and simplified navigation. Meanwhile, a wearable device used for health monitoring could provide haptic feedback for users with visual impairments.

The need for inclusive and accessible AIoT solutions is especially critical in educational contexts. Smart classrooms equipped with AI can offer personalized learning experiences, helping students with different learning needs to excel. For example, an AI-driven curriculum might adapt the difficulty level of math problems based on the student's proficiency, while also providing additional visual and auditory aids for those who require them.

Accessibility features can also be integrated into smart city solutions. Public transportation systems enhanced by AIoT can provide real-time updates in multiple formats, catering to the needs of all commuters, including those with disabilities. An AI-driven app might offer voice directions and haptic alerts to visually impaired pedestrians, ensuring their safe navigation through busy urban environments.

Incorporating usability and accessibility features from the ground up can also lead to innovations that benefit everyone, not just those with disabilities. For example, simplifying an interface for accessibility can result in a more streamlined experience that benefits all users. This principle of universal design enhances the overall appeal and effectiveness of AIoT solutions.

AIoT holds a transformative potential to make life easier, safer, and more enjoyable for everyone. The ultimate aim should be to ensure that no one is left behind in this technological revolution. By focusing on both usability and accessibility, designers and developers can pave the way for a more inclusive future where technology adapts to meet the diverse needs of all users.

In conclusion, prioritizing usability and accessibility in AIoT is not just a design preference; it's a necessity. This approach not only ensures that we build technology that works for everyone but also drives innovation by addressing a broad spectrum of user needs. As we move forward, continued focus on these principles will be essential to fully harness the potential of AIoT in enhancing user experiences across all aspects of life.

CONCLUSION

As we draw the curtain on this exploration of the revolutionary fusion of AI and IoT, it's evident that we've only scratched the surface of what these technologies can achieve together. From transforming our homes into smart sanctuaries to redefining urban living, healthcare, agriculture, and various industries, the blend of artificial intelligence and the Internet of Things has proven to be a game-changing duo. Throughout this book, we've journeyed through the intricate weaving of AI and IoT, witnessing their collective impact on our past, present, and an exciting glimpse into the future.

The most significant takeaway is perhaps the transformative potential that lies in the convergence of AI and IoT. These technologies are not just enhancing existing systems; they're fundamentally redesigning how we interact with our environment, automate mundane tasks, make strategic business decisions, and even how we perceive the possibilities of technological innovation. Smart homes are becoming the norm, cities are getting intelligent, and industries across the board are embracing automation and predictive analytics to drive efficiency and innovation.

Consider the immense strides in healthcare, where AI and IoT enable personalized medicine, remote patient monitoring, and advanced diagnostics. Patients receive timely, accurate care, often in the comfort of their homes, minimizing the need for hospital visits. Wearables and sensors track vital signs continuously, feeding data to AI systems that can alert caregivers to potential issues before they become critical. The result is a healthcare system that's not only more efficient but also more compassionate and responsive.

The industrial sector, too, has seen considerable benefits. Factories equipped with IoT sensors and AI analytics can predict equipment failures before they happen, optimize supply chains, and improve safety protocols. These benefits trickle down to consumers, who enjoy faster, more reliable service, and to companies, who gain competitive advantages through enhanced operational efficiencies and cost savings.

Yet, it's not just about efficiency and cost reduction. AI and IoT are sparking creativity and enabling entirely new business models. Retailers use AI to improve customer experiences, predicting needs before customers even articulate them. Farmers employ precision agriculture techniques, smart irrigation systems, and drone monitoring to boost yields sustainably. This illustrates a shift towards more thoughtful and effective resource management, prioritizing both economic growth and environmental stewardship.

Of course, the ascent of AI and IoT isn't without its challenges. Ethical considerations, data security, and privacy concerns are paramount. As users, developers, and policymakers, we must navigate these murky waters with care. Ensuring ethical AI practices, robust cybersecurity measures, and respect for privacy are non-negotiable pillars that should underpin every AIoT endeavor. We're reminded constantly that innovation should not come at the cost of ethical integrity. This balance will determine the sustainable success of AI and IoT integration.

Looking to the future, the horizon is bright with potential. Emerging trends like 5G, edge computing, and more advanced machine learning models will further amplify the capabilities of AI and IoT. These advancements promise to make systems faster, smarter, and more reliable, opening doors to applications we can only imagine today. The innovation ecosystem will continue to evolve, with developers, industries, and governments playing critical roles in shaping the trajectory of AIoT technologies.

What this book aims to convey is a sense of empowerment. As individuals, businesses, and societies, we hold the reins to harness AI and IoT for the greater good. Embracing these technologies allows us to address some of the most pressing challenges of our time, from climate change to healthcare accessibility. Moreover, it drives home the idea that, while these technologies are powerful, they are tools to be wielded with wisdom and responsibility.

Engagement, collaboration, and continuous learning are essential as we forge ahead. The journey of AI and IoT is one of constant evolution, requiring us to stay informed, adaptable, and proactive. The fusion of these two domains showcases the beauty of interdisciplinary innovation, where different fields converge to create something profoundly impactful.

In conclusion, the integration of AI and IoT marks a pivotal chapter in the annals of technological progress. It's a journey filled with challenges and opportunities, where the only constant is change. As we step into this brave new world, it's our collective responsibility to steer these advancements in a direction that fosters inclusivity, security, and sustainability.

May this book serve as both a guide and an inspiration as we navigate the expansive landscape of AI and IoT. The future is here, brimming with possibilities. It's up to us to shape it with ingenuity, integrity, and a vision for a better world.

APPENDIX A:
APPENDIX

In this appendix, we've compiled several useful resources and supplementary materials to enhance your understanding of the integration of Artificial Intelligence (AI) and the Internet of Things (IoT). Whether you're a beginner or an experienced professional, these additional sections aim to provide insights, tools, and references that can support your journey in exploring the transformative potential of AI and IoT technologies.

Further Reading

For those interested in deepening their knowledge, we recommend the following books and articles. These resources cover various aspects of AI, IoT, and their convergence, offering comprehensive insights:

- *The Fourth Industrial Revolution* by Klaus Schwab

- *Artificial Intelligence: A Guide for Thinking Humans* by Melanie Mitchell

- *IoT Inc: How Your Company Can Use the Internet of Things to Win in the Outcome Economy* by Bruce Sinclair

- *Artificial Intelligence and the Internet of Things: Challenges and Opportunities* - IEEE Journals

- *Smart Cities: Big Data, Civic Hackers, and the Quest for a New Utopia* by Anthony M. Townsend

Resources and Tools

To help you explore AI and IoT further, we've curated a list of online resources and tools. These platforms provide valuable information, tutorials, and software to aid in your projects and research:

- *Kaggle*: A platform for data science competitions and datasets.

- *IoT For All*: An extensive resource for IoT-related news, case studies, and insights.

- *TensorFlow*: An open-source machine learning framework by Google.

- *GitHub*: A repository for collaborative coding and open-source projects, including numerous AI and IoT initiatives.

- *OpenAI*: An AI research lab providing resources and tools for machine learning experiments.

Acknowledgments

Creating this book required the collaboration and expertise of numerous individuals and organizations. We extend our heartfelt gratitude to researchers, developers, and industry professionals who have contributed their insights and knowledge on AI and IoT. Special thanks to those who shared their success stories and case studies, providing practical examples of AIoT's transformative impact.

Contact Information

If you have any questions, feedback, or so, contact us:

Email: info@example.com

Website: www.example.com

Social Media: Follow us on Twitter, LinkedIn, and Facebook

We hope this appendix provides valuable support as you delve deeper into the fascinating world of AI and IoT. Keep exploring, innovating, and contributing to the future of technology.

Glossary of Terms

This glossary contains key terminology related to the integration of AI and IoT. Each term is defined to help you better understand the concepts discussed throughout the book.

AIoT (Artificial Intelligence of Things)

The convergence of Artificial Intelligence (AI) and the Internet of Things (IoT), where AI drives intelligent data processing and decision-making capabilities in IoT devices.

Algorithm

A set of rules or instructions given to an AI system to help it learn how to solve problems or perform tasks.

Artificial Intelligence (AI)

The simulation of human intelligence in machines designed to think, learn, and perform tasks traditionally requiring human intelligence.

Autonomous Vehicles

Vehicles equipped with AI and IoT technologies that allow them to navigate and operate without human intervention.

Big Data

Extremely large datasets that are analyzed computationally to reveal patterns, trends, and associations, particularly in relation to human behavior and interactions.

Cybersecurity

Measures and practices designed to protect networks, devices, programs, and data from attack, damage, or unauthorized access.

Deep Learning

A subset of machine learning involving neural networks with many layers that enable the modeling of complex patterns in large data sets.

Edge Computing

Data processing that occurs at the edge of a network, closer to the data source, to reduce latency and improve processing efficiency.

Home Automation

The use of AI and IoT devices within a home to automate and control various household systems, such as lighting, heating, and security.

Internet of Things (IoT)

A network of interconnected devices that collect, share, and act on data from their environment, often without human intervention.

Machine Learning

A type of AI that allows computers to learn from and make decisions based on data without being explicitly programmed for each task.

Natural Language Processing (NLP)

The ability of AI systems to understand and interpret human language as it is spoken or written, facilitating interactions between computers and humans in natural language.

Neural Network

A computing system inspired by the human brain's network of neurons, designed to recognize patterns and solve problems through learning.

Predictive Maintenance

AI-driven techniques used to predict when maintenance is needed on equipment and machinery before failure occurs, improving efficiency and reducing downtime.

Smart Grid

An electricity network equipped with IoT technology that gathers and analyzes real-time data to enhance the efficiency, reliability, and sustainability of electric power distribution.

Wearables

Iot devices worn on the body, such as smartwatches or fitness trackers, often used for health monitoring and data collection.

This glossary will help you navigate the technological landscape discussed herein, offering clarity on complex terms and concepts as we explore the transformative potential of AI and IoT.

Further Reading

To gain a deeper understanding of the intricate concepts discussed in the Glossary of Terms, there are numerous resources that can expand your knowledge and provide additional perspectives. These readings can help solidify your comprehension of the complex interplay between AI and IoT, enhance your grasp of technical jargon, and inspire you to explore the myriad opportunities and challenges in this rapidly evolving field.

Firstly, for those interested in the foundational theories and algorithms of artificial intelligence, classic texts such as "Artificial Intelligence: A Modern Approach" by Stuart Russell and Peter Norvig remain indispensable. This book covers a broad spectrum of AI principles, from basic concepts to advanced applications, making it an excellent starting point for anyone keen to understand AI from the ground up.

In the realm of IoT, "Building the Internet of Things: Implement New Business Models, Disrupt Competitors, Transform Your Industry" by Maciej Kranz offers a comprehensive overview of IoT technologies and their transformative potential in various industries. Kranz's insights into how IoT can drive business innovation and efficiency make this a critical read for professionals looking to integrate IoT into their operations.

For a more specialized focus on the convergence of AI and IoT, "AI and IoT: The Internet of Things Converging with Artificial Intelligence" by Jayden Wallace and Solomon Peyton explores how these technologies intersect and complement each other. This book delves into practical applications, case studies, and strategies for leveraging AI to enhance IoT functionalities in real-world scenarios.

Additional insights on AI can be sourced from "Deep Learning" by Ian Goodfellow, Yoshua Bengio, and Aaron Courville. This work delves into the intricacies of neural networks and deep learning techniques, providing readers with an in-depth exploration of one of the most impactful AI advancements.

For those with a specific interest in the ethical and societal implications of AI and IoT integration, "Weapons of Math Destruction" by Cathy O'Neil is a must-read. O'Neil argues how algorithms can propagate bias and inequity, thus stressing the importance of ethical considerations in the deployment of these technologies.

Understanding the regulatory landscape is crucial for navigating the legal challenges surrounding AI and IoT. Books like "The Law of Artificial Intelligence and Smart Machines: Understanding AI and the Legal Impact" edited by Theodore F. Claypoole offer valuable insights into the current regulations and emerging legal frameworks governing AI and IoT applications.

To grasp the economic and strategic dimensions of deploying AIoT solutions, "Artificial Intelligence for Business: A Roadmap for Getting Started with AI" by Doug Rose is an excellent resource. Rose's pragmatic approach helps businesses identify AI opportunities, develop strategic roadmaps, and measure the impact of AI implementations.

For practical guidance on implementation, "IoT Inc: How Your Company Can Use the Internet of Things to Win in the Outcome Economy" by Bruce Sinclair provides actionable advice on how to successfully embark on IoT projects, from conceptualization to full-scale deployment.

Another noteworthy read is "Predictive Analytics: The Power to Predict Who Will Click, Buy, Lie, or Die" by Eric Siegel. This book explains how predictive analytics, an essential component of AI, can transform data into decision-making tools, illustrating its potential through a variety of real-world examples.

Open source publications and research papers are also invaluable for staying abreast of the latest advancements. Websites like ArXiv.org and Google Scholar host a wealth of peer-reviewed articles and preprints on cutting-edge AI and IoT research topics. These sources are particularly beneficial for those interested in the theoretical underpinnings and future trajectories of these technologies.

For those who prefer digital and multimedia content, online courses from platforms like Coursera, edX, and Udacity offer a wide array of lessons on AI and IoT. Courses such as "AI For Everyone" by Andrew Ng on Coursera demystify AI concepts for a broad audience, while more technical courses dive into specific algorithms and applications.

Engaging with academic journals can also widen one's understanding. Journals like the IEEE Internet of Things Journal or the Journal of Artificial Intelligence Research publish the latest scientific findings and are excellent resources for those wanting to delve deeper into professional and academic discourse.

Attending conferences and symposiums can provide firsthand learning opportunities from leading experts. Annual events such as the AI Summit, IoT Solutions World Congress, and the NeurIPS Conference offer a platform for networking, learning emerging trends, and discovering innovative applications of AI and IoT.

Lastly, joining professional organizations like the Association for Computing Machinery (ACM), the Institute of Electrical and Electronics Engineers (IEEE), or the International Association for Artificial Intelligence (IAAI) offers access to exclusive resources, workshops, webinars, and communities dedicated to fostering knowledge and innovation in AI and IoT.

Leveraging these diverse resources will not only deepen your understanding of the terms and concepts discussed in this glossary but also position you to effectively navigate and contribute to the transformative world of AIoT. As you broaden your horizons through further reading, you'll be better equipped to harness the full potential of these powerful technologies.

Resources and Tools

Leveraging the various resources and tools available is crucial for anyone delving into the realm of AI and IoT. Whether you're a seasoned professional or a newcomer, these resources provide the foundation and advanced knowledge needed to understand and effectively implement AIoT solutions. Here, we'll explore different types of tools and resources that can help you navigate this burgeoning field.

Firstly, comprehensive documentation and tutorials are essential starting points. Many AI and IoT platforms offer extensive documentation that details how to get started, best practices, and troubleshooting tips. For instance, platforms like TensorFlow and PyTorch provide a wealth of tutorials that cover both fundamental and advanced topics in AI. Similarly, IoT platforms like Arduino and Raspberry Pi come with extensive guides that walk users through initial setups to more complex integrations.

Another vital resource is the vast array of online courses and certifications available. Platforms such as Coursera, edX, and Udacity offer courses tailored to different skill levels. These courses often include hands-on projects that enable learners to apply their knowledge practically. Certifying in specialized tracks like Google's AI and IoT certifications or IBM's Data Science Professional Certificate can also add significant value to your credentials.

Books and research papers also continue to be indispensable sources of in-depth knowledge. Publications from leading experts and institutions provide comprehensive insights and case studies. Books like "Artificial Intelligence: A Modern Approach" and "Data Science for IoT Engineers" serve as excellent references for both theoretical knowledge and practical applications. Moreover, keeping abreast of the latest research papers via platforms like arXiv and Google Scholar can help you stay updated with the most recent advancements and innovations in AIoT.

Forums and community support networks are invaluable, especially when facing specific challenges or seeking advice on best practices. Websites like Stack Overflow and GitHub offer communities of experts who can assist with troubleshooting and share code snippets, project ideas, and other resources. Participating in these forums can also provide networking opportunities, enabling collaboration with others working on similar projects.

The importance of software development kits (SDKs) and application programming interfaces (APIs) can't be overstated. SDKs and APIs simplify the integration of complex functionalities into your projects. For instance, Microsoft's Azure IoT SDK and Google's Cloud IoT Core provide robust tools for building and scaling IoT solutions. Similarly, AI toolkits like OpenCV for computer vision and NLTK for natural language processing streamline the implementation of sophisticated AI capabilities.

Simulation tools offer another layer of valuable support, particularly in the testing phase. Tools like MATLAB and Simulink allow for the simulation of complex systems before deployment. These tools help identify potential issues and optimize performance, ensuring smoother real-world implementations. IoT-specific simulation platforms such as ThingSpeak enable the visualization of sensor data and the testing of IoT algorithms in a controlled environment.

Data is the backbone of any AI and IoT application, making access to quality datasets a critical resource. Platforms like Kaggle not only offer datasets but also host competitions that drive innovation. Public datasets from organizations like NASA, WHO, and various government agencies provide reliable data for research and application development. Curating high-quality data is also essential, so tools for data cleaning and preprocessing, like Pandas and Scikit-learn in Python, are equally crucial.

Visualization tools play an important role in understanding and presenting data. Tools like Tableau, Power BI, and Plotly offer interactive dashboards and visualization capabilities that can turn complex data into actionable insights. These tools are not just for data scientists but are equally important for stakeholders who need to make informed decisions based on AIoT metrics.

Open-source frameworks and libraries empower developers to build robust AI and IoT solutions. Libraries such as TensorFlow, Keras, and Scikit-learn for AI, and Node-RED and Johnny-Five for IoT, offer a wealth of pre-built functionalities and a large support community. These tools accelerate development time and foster innovation by providing scalable and customizable solutions.

Cloud services have revolutionized the way AI and IoT applications are developed and deployed. Platforms like AWS IoT, Google Cloud IoT, and Microsoft Azure IoT Hub offer scalable cloud solutions that handle data processing, storage, and analysis. These platforms provide the necessary infrastructure to support large-scale IoT deployments and the integration of advanced AI models.

Lastly, security tools are indispensable in the AIoT space. With the integration of numerous devices and extensive use of data, ensuring security is paramount. Tools like Wireshark, Nessus, and Snort provide advanced capabilities for monitoring network traffic, detecting vulnerabilities, and ensuring the integrity of IoT devices. Employing these security measures protects against potential breaches and ensures the reliability of your AIoT applications.

Charlie Morgan

In summary, a rich array of resources and tools is available to support your journey in the AI and IoT landscape. From documentation and tutorials to cloud services and security tools, these resources empower you to develop, deploy, and maintain effective AIoT solutions. Leveraging the right mix of these tools can significantly enhance your ability to innovate and drive meaningful change in various industries.